S0-BWY-398

A Guide To Biting and Stinging Insects and Other Arthropods

Charles S. Papp
and
Lester A. Swan

Second enlarged edition
with
more than 300 illustrations

ENTOMOGRAPHY PUBLICATIONS

Second enlarged edition
First edition: 1981

ISBN 0-9608404-0-0

ENTOMOGRAPHY PUBLICATIONS
1722 J Street, Suite 19, Sacramento, CA 95814
(916) 444-9133

A product of Sierra Graphics and Typography

TABLE OF CONTENTS

Preface **5**

Introduction **7**

The Stingers — 7
The Biters — 8

I. BEES **10**

Bumble Bees — 11
The Honey Bee — 15
Squash Bees — 17
Sweat Bees — 19
Leafcutting Bees — 21
Importance of Bees — 23
The Bee Sting — 24
First Aid and Prevention — 26
The African Honey Bee — 28

II. WASPS **29**

Paper Nests — 31
Yellowjackets and Hornets — 33
Paper Wasps — 35
Potter and Mason Wasps — 36
Spider Wasps — 37
Mud Daubers — 39
Sand Wasps — 40
Parasitic Wasps — 41
An Ounce of Prevention — 42
Wasp Stings — 43
Removal of Paper Nests — 44

III. ANTS **46**

Legionary or Army Ants — 48
Harvester Ants and Others — 50
Fire Ants — 56
Carpenter Ants — 58
Ant Stings — 59
First Aid and Prevention — 61

IV. FLIES **62**

Biting Midges — 62
Black Flies — 64
Horse Flies and Deer Flies — 70
Stable Flies and Close Relatives — 73
Snipe Flies — 75
Eye Gnats and Louse Flies — 76
Mosquitoes — 77
Nature of Fly Bites — 85

V. FLEAS AND SUCKING LICE **87**

Fleas — 87
Nature of Flea Bites — 91
Sucking Lice — 92
Nature of Lice Bites — 94

VI. BUGS AND BEETLES **96**

Assassin Bugs — 95
Water Bugs — 98
Bed Bugs — 100
Nature of Bug Bites — 101
Beetles — 101
Tiger Beetles — 102
Charcoal Beetles — 103
Blister Beetles — 103
Nature of Blister Beetle Injury — 106
Stinging Beetle Larvae — 106
Nature of Beetle Larvae Stings — 107
Darkling Beetles — 108

VII. STINGING CATERPILLARS **110**

Giant Silkworm Moths — 111
Tussock Moths — 116
Tiger Moths — 117
Some Other Stinging Caterpillars — 118
Tent Caterpillars — 120
Caterpillar Stings — 121
First Aid and Prevention — 123

VIII. MITES AND TICKS **125**

Mites — 127
Ticks — 132
Spotted Fever and Tick Paralysis — 138
First Aid and Prevention — 139
Repellents — 140
Summary (Table) — 142-146

IX. SPIDERS, SCORPIONS AND CENTIPEDES **147**

Whipscorpions, False Scorpions and Windscorpions — 147
Spiders — 150
Tarantulas — 153
The Harmful Spiders — 154
Black Widow Spiders — 154
Violin Spiders — 156
Spider Bites: First Aid and Prevention — 157
Scorpions — 159
Scorpion Stings: First Aid and Prevention — 163
Centipedes — 165

Summary of Precautions Against Arthropods That Attack Man **167**

First Aid **168**

Appendices **169**

Appendix 1 (The Plague) — 170
Appendix 2 (Fleas) — 175
Appendix 3 (Rabies) — 184

Glossary **191**

Bibliography **200**

Index to Illustrations, Common Names, and Scientific Names **205**

PREFACE

The purpose of this book is to acquaint the reader with the common insects and other arthropods (spiders, mites, ticks, scorpions and centipedes) that sting or bite man, so he may guard against them and know how to best deal with them when a painful encounter is unavoidable. While these stings and bites are generally more annoying than harmful, and have no more serious consequences than an itching and more or less painful swelling in the majority of cases, they can have serious consequences in many instances. The itching and swelling are caused by toxins (venom) injected during the act of stinging or biting. People react differently to these substances and have varying degrees of sensitivity to them; for some the reaction is slight, for others it may be violent and even result in death. There are in fact many cases on record (and probably many more not recorded) of fatalities resulting from the sting of the common honey bee and the familiar wasps.

The book is non-technical in nature and intended for the general reader, who will require no previous knowledge of the subject. It should be of value to hunters, campers, fishermen, hikers, boy scouts and all who venture into the great outdoors, and as a "first aid" reference in the home, especially in the sprawling and ever-growing suburbia where one is "closer to nature" than in the cities. It is hoped that the book will also be useful to doctors and to students or workers in the field as an introduction or handy reference. A recognition of the arthropods dealt with here and some of the knowledge of their habits is the first step toward prevention of painful stings and bites. If they can't be avoided, identification of the offenders will indicate the proper action to take; when the reaction is severe, a physician should be consulted (in the extreme cases prompt action may be necessary to avoid fatal consequences). Proper identification in this case will lead to more prompt and effective treatment; it is important to get a specimen of the arthropod involved (dead or alive) if possible, and to save it to confirm your own identification or for help in making one.

Generally one can distinguish a "bite" from a "sting" by the sharp and burning sensation of the latter. Only the bees, wasps, ants and scorpions sting *in the strict sense; the others* bite. *Honey bees usually leave their barbed stingers imbedded in the flesh and prompt removal by* scraping *with a sharp knife (or fingernail) will avoid a prolonged reaction from the sting. For bites and stings in general the application of an antiseptic will reduce the possibility of infection. Most people will find a dab of ammonia an effective first aid measure for the relief of a bite or sting, and swabbing the area with rubbing alcohol or an alcoholic solution of witch hazel may help. Ice packs or cold compresses applied promptly will bring relief and prevent the rapid spread of venom throughout the blood stream; this and the application of a tourniquet (between the wound and heart) are recommended in the case of severe stings prior to medical treatment.*

Charles S. Papp
Lester A. Swan

Preface to the Second Edition

On the requests and comments of the first edition of this handy reference book — designed as an aid for the average out-of-doors American — there are 37 new pages added with valuable information.

Three "Appendices" are now included which are especially useful for those who more frequently enjoy Nature and have more opportunities to come in contact with the various creatures that may carry disease, including rabies.

Several illustrations are not cited in the text. These are presented to help identify some of the insects and their relatives; many of these are very closely related to the harmful species, or may seem dangerous because of their awsome appearance, yet be just simple inhabitants of nature.

It is hoped that this new edition may give you a more detailed view into the lives of Nature's little creatures and may help you to avoid any serious consequences while in contact with them.

Sacramento, California — *Charles S. Papp*

NOTE: Numbers in parenthesis in the text refer to references cited. See Bibliography on page 200.

INTRODUCTION

To get the matter straight, without becoming too involved in classification, we need only point out that spiders, mites, ticks and scorpions are not insects, strictly speaking. They are *arachnids* (class Arachnida) and differ from insects (class Insecta) in having four pairs of legs instead of three in the adult stage, and no more than two body parts (cephalothorax and abdomen) instead of three (head, thorax and abdomen) as found in insects; they also lack the antennae characteristic of insects. Nor are centipedes insects in the strict sense; they are *chilopods* (class Chilopoda) and differ in having a pair of legs on most body segments. Arachnids and chilopods never have wings as most insects do. All three classes belong to a single group called *arthropods* (phylum Arthropoda), which are characterized by their jointed legs. (89)

The Stingers

Among the insects, only the group (order) called hymenoptera — characterized by two pairs of membranous wings, and comprising the bees, wasps and ants — have a stinger, which is located at the tip of the abdomen of the females only. The stinger is strictly a defensive weapon, and an evolutionary modification of the ovipositor — an egg-laying instrument resembling a hypodermic needle which is possessed by the females of hymenopterous parasites for placing eggs in their hosts. [The ovipositor is often used by female parasitic wasps for stinging and paralyzing the host temporarily and in some cases permanently and sometimes for sucking up the liquid body contents of the host. Some of the larger parasitic wasps called ichneumons have been known to sting collectors, with very painful results for the latter.] Yellowjackets and hornets are the most troublesome stingers among the wasps, and pose a threat with their presence near and around houses. Ants can sting *and* bite; fortunately not all of them sting nor do all of them bite. The stinging trait is found mostly among the so-called army ants, fire ants and harvester ants.

Most of the honey bee varieties used in apiaries today are normally gentle and will attack only when provoked or irritated. In trying to remove her stinger the bee literally tears it away from her body and thereby signs her own death warrant. It is well to remember that when the bee stings, you are apt to be pounced upon by the others that have responded to the warning given by the first bee. "The odor from the sting left behind by the stinging bee serves as a target marker for subsequent defensive attacks . . . The angry bees tend to sting at the same spot" (10). Other bees and wasps do not have a barbed stinger and can sting with impunity more than once. Bumble bees are normally quite gentle and aroused only when the nest is stepped on or otherwise disturbed. An aroused swarm is not to be dismissed lightly, as anyone who has been caught in this predicament can testify.

The stinger possessed by scorpions is a sharp, curved instrument at the end of the tail and resembles a spider's fang somewhat. Their stings are very painful and the venom in some species is very toxic. The so-called stinging caterpillars, which are the larvae of certain moths and butterflies, are of course "stingers" of another sort. These caterpillars have urticating hairs — specialized protective hairs with stinging properties that can cause an irritating skin rash. Here again the sensitivity of different persons varies, and the degree of irritation varies with the species of caterpillar. Children are intrigued by their gay colors or fuzzy appearance and are tempted to handle them.

The Biters

The majority of insects and other arthropods that we are dealing with here are biters, that is, they use their mouthparts to inflict injury. Some do this with their mandibles or jaws, which may be toothed or curved and sharply pointed; beetles and ants are examples. Others bite by means of modified mouthparts which are adapted for piercing and sucking, as for example, the true bugs, mosquitoes, fleas, lice, mites and ticks. Aside from their annoyance as biters, many of them are vectors or carriers of disease organisms affecting man. Spiders bite with hollow, sharply pointed fangs attached to the chelicerae or "jaws." Most of them are venemous but not many will normally bite since they are too timid; only the black widow and brown recluse spiders are considered dangerous. Centipedes bite with their claws, which are modified appendages of the first body segment; the large western species found under rocks and logs can inflict a painful bite.

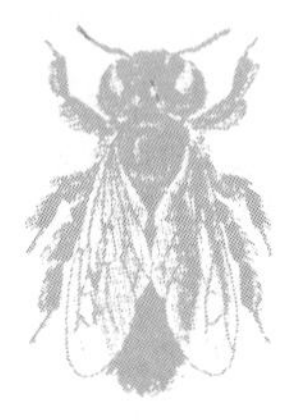

I. BEES

Bees are found wherever flowers grow and suitable nesting sites are available. They belong to the great superfamily Apoidea (order Hymenoptera). Of the 3,300 species and subspecies recorded for North America north of Mexico, more than 2,500 are solitary bees, and about 400 are social or semisocial; the remainder are parasitic in the nests of other bees (51). Solitary bees, as suggested by the name, nest individually. The females of ground-nesting species dig tunnels in the ground, usually shaft with several branches each ending in a cell, which is provisioned with pollen and usually some nectar, and sealed after an egg is deposited. Others make nests by hollowing out a plant stem, tunnelling wood, or using ready-made cavities, and dividing these into a series of cells, each of which is stored with pollen and nectar and implanted with an egg. The solitary bees are most abundant in warm arid regions as found in western North America. In places of habitation they are greatly outnumbered by honey bees, or at least are not nearly so conspicuous. Most of them are small and not likely to sting, or if they do it would be no more severe than a pinprick in most cases.

There are two kinds of social bees in North America north of Mexico, the honey bee and the bumble bee (the stingless bees, which are also social, are tropical). By "social" is meant, the bees live in organized colonies, with a division of labor in the form of a caste system. [Besides bees, paper wasps and ants, the termites, though more primitive, are also social in behavior, resembling the ants in this respect. While not strictly "social," some other insects care for their young in other ways, as will be seen later in dealing with solitary bees and wasps. Similar concern for the young is shown in still other ways by some other insects, as for example, the earwig, a shiny brown insect with forceps-like tail which occasionally shows up in garbage cans. The female broods her eggs which are laid in a nest in the ground, and the newly hatched young are guarded by both parents and prevented from escaping for a few days until ready to face the world on their own. A tabanid fly, *Goniops chrysocoma,* a large smoky brown eastern species with a small head and dense coat of golden brown hairs, also broods her eggs, which are laid on the underside of leaves in wooded areas; she remains with them until they are hatched and for several hours after, when her mission is completed and she dies.] There are three distinct castes among social bees: the queens — female reproductives whose function is to mate and lay eggs; the drones — male reproductives whose sole function is to provide sperm for the fertilization of eggs; and the workers — females who do all the work of the colony: constructing cells, gathering nectar and pollen, feeding the queen, caring for and feeding the larvae, cleaning and guarding the hive. Thus, it is the worker that we commonly see. They may be distinguished by their size. The queen is the longest, the drone is somewhat shorter but stouter, the worker is the smallest.

Bumble Bees

Bumble bees, (family Apidae) are more primitive than honey bees, being less "advanced" or complex in behavior, as shown in the construction of nests, size of colonies, their manner of bridging the seasons, and in other forms of behavior. They are the dominant bees in the higher latitudes (60 to 65 degrees) of Canada and Alaska. There are about 50 species and numerous subspecies of bumble bees in North America north of Mexico. They are usually black and yellow and stouter than honey bees; they look much larger partly because of the dense coat of hairs. The very big ones seen in the spring are queens busy building new nests and preparing for their first brood; when the latter — usually all workers — emerge as adults, they take over the chores of the hive and the queen confines herself to laying eggs. [The queen bee (as does the female of many other Hymenoptera) regulates the sex of her progeny. When she mates (and this is the case with most other insects), the sperm is stored in the spermatheca, a kind of pouch. When the queen bee lays her eggs, however, she releases sperm to fertilize only those eggs that are to become females and does not fertilize those that are to become males or drones. Females selected to become queens are given a special diet; females not so fed become workers.]

Bumble bee nests are made of grass on or under the ground — under rocks, in a large clump of grass, or in an abandoned mouse or bird nest. Brood cells, in which the eggs are laid and larvae develop, and honey pots for storage of honey are made of pollen and wax. The queen starts with a few cells — usually not more than eight or ten in all — to get the colony going, and as new broods of workers arrive it is gradually expanded. Brood cells are not used again but cocoons, when empty and after altering, are used for storing pollen and honey. Broods of larvae that are to become queens or males, and the first larval stage of those destined to become workers, are fed a regurgitated mixture of honey and pollen. Bumble bee hives are small compared to those of the honey bee, probably containing on the average no more than 200 bees. The colony breaks up in the fall; only young mated queens hibernate and survive the winter.

The **Nevada Bumble Bee** *(Megabombus nevadensis)* is found throughout the western United States and Canada, eastward in the middlewestern states to Wisconsin and Illinois, and southward from New Mexico into Mexico; it is black, with intermixture of yellow on face and top of head, yellow on thorax, yellow band on abdomen, wings clouded chocolate-brown, a large species, from 17 mm (workers) to 22 mm (queens) long, male intermediate in size, some specimens from the lower coastal Northwest grading into almost entirely black. The **Yellow Bumble Bee** *(Megabombus fervidus)* is widespread throughout the United States and Canada, except the Pacific coastal area, predominantly yellow, with intermixture of black, head black, wings clouded with black, from 12.5 to 22 mm long; the subspecies *californicus* is strictly western, and is highly variable in the amount of yellow. *Pyrobombus huntii* occurs from the southwestern states northward to Montana, Alberta and British Columbia; it is black and yellow, about the same length as *M. fervidus* but more robust.

Pyrobombus griseocollis extends from the eastern part of the United States west across the northern states to the Pacific coast, and from Washington to northern California; it is predominantly black, with varying amounts of brownish yellow intermixed on head, thorax and abdomen, wings clouded with reddish brown to blackish, from 13 to 21 mm long. *Pyrobombus morrisoni* is largely western, extending from the midcontinent

Fig. 1. Some species of bumble bees: A — *Megabombus nevadensis;* B — *Megabombus fervidus;* C — *Pyrobombus huntii;* D — *Pyrobombus griseocollis* (UDSA photo).

Fig. 2. Nest of a bumble bee *(Pyrobombus morrisoni):* A — honey pots; B — pollen tube; egg baskets on cocoons (one opened); C-D — young brood in wax cells (USDA photo).

to the Pacific coast, and from British Columbia to Mexico; it is mostly black, with most of head and part of the thorax golden yellow, a large species, from 17.5 to 22 mm long. The **Yellow-Faced Bumble Bee** *(Pyrobombus vosnesenskii)* occurs along the Pacific coast, east to Colorado; it is black, with face, top of head, pronotum, mesonotum in front of wings, and one segment of abdomen yellow, from 10 to 20 mm long, one of the commonest bumble bees in many parts of California, often mistaken for the **Yellow Bumble Bee.**

Bumble bee nests are often invaded by cuckoo bees of the closely related genus *Psithyrus*. The parasitic cuckoo bees have no worker caste and lack the adaptation for carrying pollen; they lay their eggs in the center of the brood mass, and the larvae (which hatch before the others) consume their food. The

Fig. 3. The honey bee *(Apis mellifera)*: A — worker; B — queen; C — drone; D — brood comb of healthy brood of bees (USDA drawings and photo).

damage of these usurpers is often considerable. Cuckoo bees are mostly black with dense yellow pile on the head and thorax. Their hum is softer than that of the bumble bee *(Psithyrus* means "whisperer"), and they are powerfully built, thus capable of stinging harder. (Cuckoo bees have no "pollen basket" or similar adaptation that pollen gathering bees have. The pollen basket is a smooth concavity fringed with hairs, on the *outside* of the tibia, second joint from the end, of the hind leg. Pollen is gathered in the "pollen brush" — rows of bristles on the *inside* of the basal segment, next to the tibia, of the tarsus or "foot" of the hind leg, the tarsus itself has five segments, the basal one being greatly enlarged in bees for this purpose. Bees rub their hind legs together to transfer the pollen from the brush of one leg to the basket on the opposite leg; it is packed into the latter and carried to the hive.) *P. insularis* is widely distributed throughout the United States and Canada, and "probably the best known western species"; the head is black with conspicuous yellow tuft between base of eyes, thorax yellow with black band between base of wings, abdomen black mostly with yellow on sides, males from 10 to 14 mm long, females from 14 to 18 mm long.

The Honey Bee

The **Honey Bee** is a distinct species, *Apris mellifera* (family Apidae) and is not a native of this continent; it was brought here by early settlers sometime before 1638 and is now worldwide in distribution. Most of the honey bees are "domesticated," but many of them have escaped and are domiciled in nests of their own apart from apiaries. There are several races of **Honey Bee** in common use in apiaries in North America: the Italian bee, which is black, with yellow bands, and gentle in nature; the Caucasian bee, which is black, with gray bands, and normally mild-mannered but disagreeable when aroused; and the Carniolian, grayish in color but otherwise resembling the Caucasian bee. Carniolians are the most gentle of the three but have the strongest tendency to swarm; they are good comb builders and produce many white cappings on the cells. The brighter color of the Italian bee and the greater ease in finding the queen have made it a popular variety. Queens are from 15 to 20 mm long, drones from 15 to 17 mm long and more robust, workers from 11 to 15 mm long.

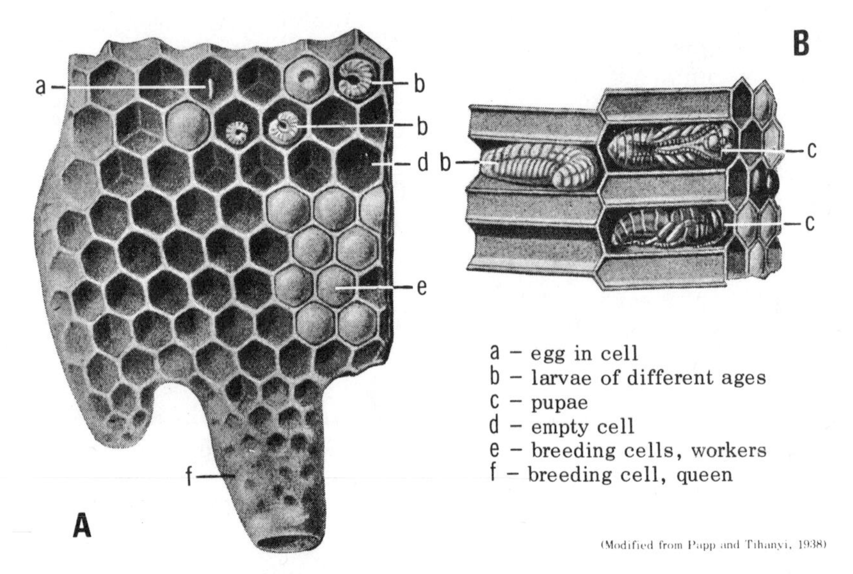

Fig. 4. Inside the honey bee comb: A — portion of comb showing large cell (f) and cells with various developmental stages of workers; B — cross section of comb showing two-layer arrangement of cells.

In the **Honey Bee** hive, wax combs composed of hexagonal cells, for rearing of the larvae and storing honey, are constructed in two horizontal rows back-to-back. Worker broods are kept in the lower central part of the comb, the larger cells of the drones are in the lower corners; honey is stored nearby in cells of the same size in each case. The large thimble-shaped queen cells hang down vertically from the brood comb. The white elongated eggs are attached upright to the base of the cells, one in each, and hatch in about three days. The brood cells are kept open during feeding of the larvae and capped when they are ready to pupate (or change to an adult). All the larvae are fed royal jelly for the first two days; thereafter, the future drones and workers are fed a mixture of honey and pollen while the future queens remain on the diet of royal jelly (and continue on it as adults). Newly emerged workers become nurses, and later produce wax; after about three weeks of various chores in the hive, they become foragers. The drones are fed until they become a burden; when the hive gets crowded and stores are no longer plentiful, they are ousted from the hive.

Swarming of honey bees takes place about the time the population and activity in the hive reach their peak and when the new queens are produced. At this time the old queen leaves, taking half of the colony with her. They come to rest in a cluster suspended from the limb of a tree or other overhang, and remain here until the scout bees have located promising new nesting sites and return with the information. After the scouts have performed their dances on the cluster, thus communicating the various possibilities to all assembled, a unanimous decision is somehow reached and the swarm flies off "in a swirling mass" to the place chosen for its new domicile. [While the mechanism is not clearly understood, the social bees are believed to communicate through "pheromones" — chemical substances produced by the bees and emitted at the proper times to bring about the coordinated response of the other members of the colony. Bees undoubtedly have a keen sense of smell and perceive colors as well — at least the basic yellow, blue-green, blue and ultraviolet, but not pure red — which apparently aids them in finding their way over distances. The many-faceted compound eye of the bee is also receptive to polarized light and can distinguish its direction, which is an aid in navigation as long as the blue sky is visible and the sun is not at its zenith. (Polarized light vibrates only in one plane, not in all planes as does the normal light we perceive.)] Meanwhile, back in the old hive the young queens engage one another in mortal combat until only one remains alive; when a week old, the survivor flies off to mate and returns the undisputed queen. Her normal life expectancy is about one year, that of a worker is about six weeks. The drones live about eight weeks on the average and die after mating. If the beekeeper provides sufficient space for brood rearing (by adding "supers" or compartments properly placed), colonies with a vigorous queen will not swarm. Unlike bumble bees, honey bees maintain their hives throughout the winter by their habit of storing honey and clustering. The activity of the bees clustered about the eggs and larvae, and their consumption of the honey stores generate sufficient heat to keep them from freezing.

Squash Bees

The squash bees (genera *Peponapis* and *Xenoglossa,* family Apidae) are large bees resembling the **Honey Bee;** they are remarkable in that the females obtain pollen only from the flowers of cucurbits (squashes, gourds, pumpkins). They are

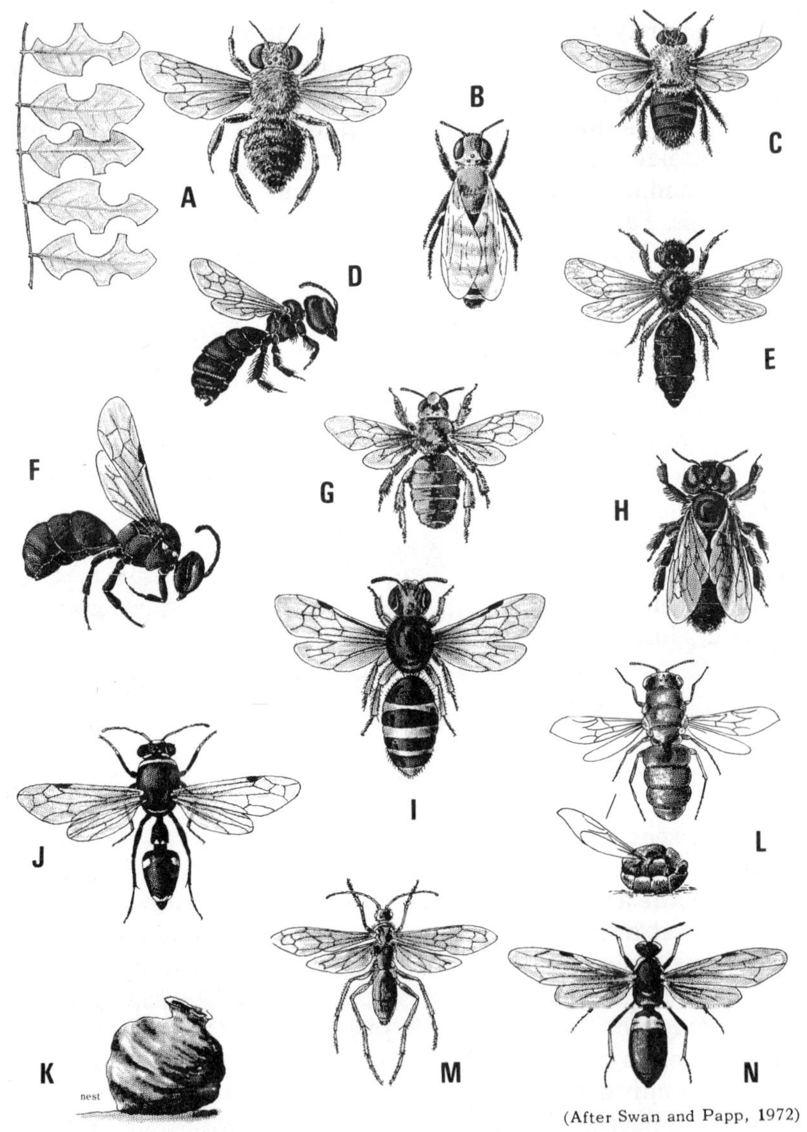

(After Swan and Papp, 1972)

Fig. 5. Some solitary bees and wasps: A — leafcutting bee *(Megachile latimanus)* and damaged leaves; B — alfalfa leafcutting bee *(Megachile rotundata)*; C — large carpenter bee *(Xylocopa virginica)*; D — little carpenter bee *(Coratina acantha)*; E — mining bee *(Andrea carlini)*; F — a mason bee *(Hylaeus modestus)*; G — a mining bee *(Anthophora occidentalis)*; H — carpenter bee *(Xylocopa tabaniformis)*; I — sweat bee *(Halictus zonulum)*; J — potter wasp *(Eumenes fraterna)*; K — mud nest of J above; L — cuckoo wasp *(Parnopes edwardsi)*; M — tarantula hawk *(Pepsis formosa)*; N — a mason wasp *(Monobia quadridens)*.

active shortly after sunrise, when these plants make their pollen available, and are often mistaken for honey bees, more especially when dusted with pollen. Another interesting thing about them is their habit of sleeping in the closed (and usually staminate) flowers, from which they emerge covered with pollen. They spend a considerable amount of time sleeping, especially the males. *Peponapis pruinosa* is transcontinental in distribution, from Maine to Georgia, west to Idaho, California and Mexico; it is black, with brownish yellow pubescence on head and thorax, and bands of whitish or yellowish hairs on the abdomen, from 12 to 15 mm long. It digs a vertical shaft in the ground with several encircling laterals, each terminating in an ovoid cell, which is provisioned with pollen and nectar and closed after an egg is placed on the mass. The entrance is surrounded by an accumulation of dirt for a short time but bee traffic soon reduces it to a funnel shape. The squash bees overwinter in the prepupal stage within a cocoon in the soil, and change to an adult and emerge the following summer. There are eleven species of *Peponapis* and *Xenoglossa* known to occur in North America north of Mexico (45).

Sweat Bees

The sweat bees (genus *Halictus,* family Halictidae) are so named for their attraction to perspiration. They frequently alight on the arms or other exposed parts of the body of sweaty persons but are not likely to sting unless swatted or inadvertently pressed against the skin. They show a strong tendency toward social behavior, sometimes resembling the bumble bees in this respect. Typically, the female, after emerging from hibernation in the spring, digs a curving burrow in the soil with cells branching off to the sides; these are provisioned with pollen and nectar and an egg laid in each. The first brood is all females, as in the case of bumble bees (see page 11); these bees build their own cells, branching off from the parental burrow or independent of it, and since there are no males the eggs are unfertilized and produce males only. The mother meanwhile continues to lay eggs on pollen balls made by her daughters. Since she has sperm (from mating the previous fall, before hibernating), she produces females again, and they now mate

Fig. 6. Female alkali bee *(Nomia melanderi)* at entrance to nest (UDSA photo).

with the males. The males die after mating, and the females hibernate, to repeat the cycle the following year.

The halictid bees comprise a large family — nearly 500 species in North America north of Mexico — with varying habits, from solitary to semisocial. A common western species, typical of the sweat bees, is *Halictus farinosus,* which ranges from California to New Mexico, northward to Nebraska, Montana, and British Columbia; it is black, with yellowish legs, reddish yellow markings on abdomen, head and thorax with dense coat of long hairs, about 10 mm long. The family includes one of the most useful bees, the **Alkali Bee** *(Nomia melanderi),* so named because it is found in the alkali soils of the far western states: Washington, Oregon, Idaho, Wyoming, Colorado, Utah, Nevada and California; it is black, and easily recognized by the light emerald-green, highly polished bands across the apical segments of the abdomen, a little smaller than the **Honey Bee.** Each female builds a nest in about 30 days, consisting of an entrance shaft (marked by a mound of dirt) with 15 to 20 cells branching off at various levels. Aggregations are sometimes very large, with numerous tunnel entrances concentrated in a small area, and the bees coming and going in

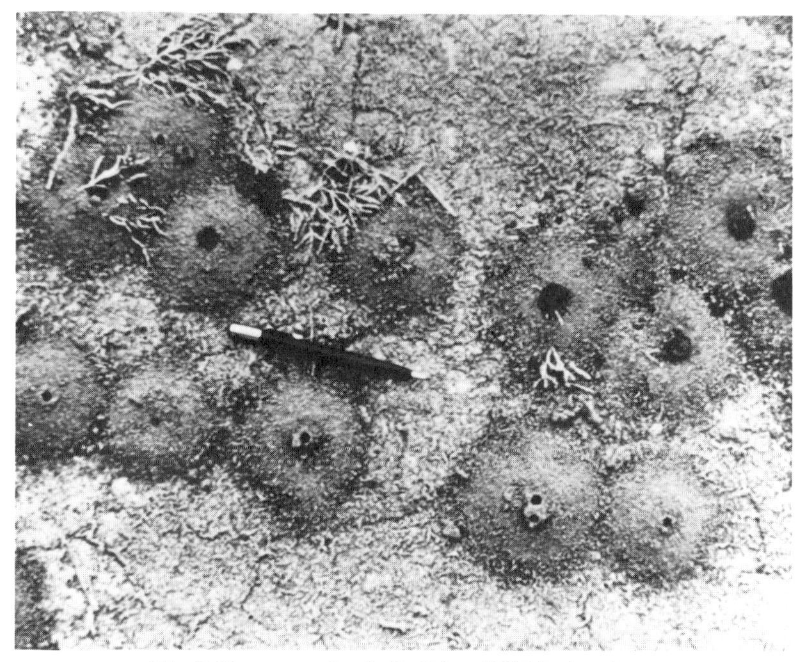

Fig. 7. Nest mounds of alkali bee (USDA photo).

busy succession during the provisioning operations. This is one of the most important pollinators of alfalfa grown for seed; growers encourage them by building large artificial nesting sites (33).

Leafcutting Bees

The leafcutting bees (genus *Megachile,* family Megachilidae) are also important pollinators of alfalfa. They are usually black with yellowish pubescence and whitish bands on several of the abdominal segments. Megachilid bees do not have a "pollen basket" on the hind legs. They have a "pollen brush" on the underside of the abdomen which is used to carry pollen back to the nest; the brush is most often reddish, but is yellowish or black in some species. The leafcutting bees comprise a very large family of solitary bees with about twice as many species as the halictids; they usually bore tunnels in wood, in the ground, or use any ready-made cavity that suits the purpose,

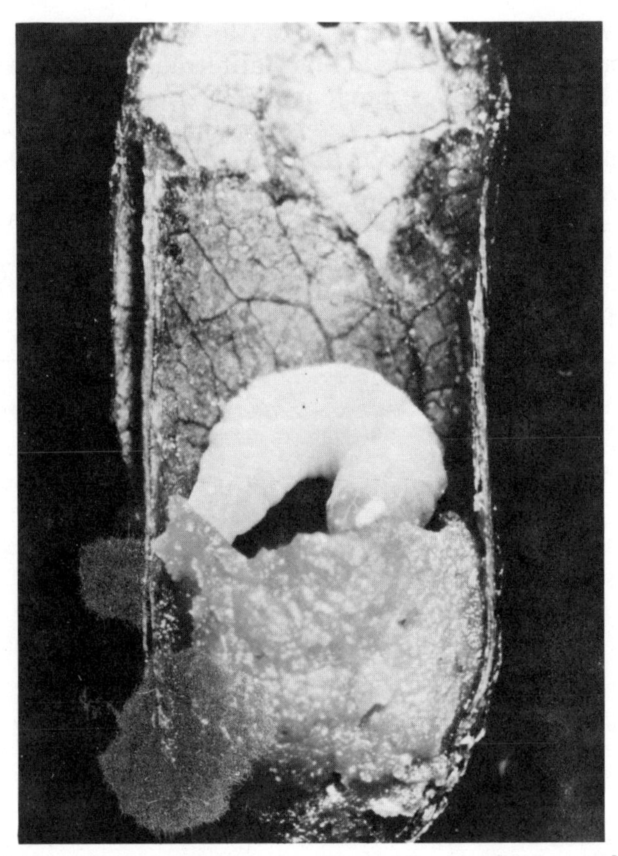

Fig. 8. Cell of alfalfa leaf cutting bee *(Megachile rotundata)* opened to show construction; also shows larva and food store (USDA photo).

and construct thimble-like cells shaped from pieces of leaves cut evenly from rose or other plants. When a cell is provisioned with pollen and nectar and an egg inserted, it is sealed over with round pieces of leaves cut a little larger than the diameter of the cell, making a tight fit. Several cells are made end to end and several tunnels may be constructed side by side.

Megachile latimanus is a widespread species ranging from Nova Scotia to Georgia, westward to Alberta and British Columbia, Wyoming and Colorado; it is blackish, densely clothed with pale brownish yellow pubescence, faint white bands on abdominal segments, pollen brush on underside of female pale red, from 12 to 15 mm long. *M. rotundata* was accidently introduced into the eastern United States in the

1930's, and has spread westward to Alberta, Oregon, Washington, Montana, South Dakota and California; our smallest leaf-cutting bee — from 6 to 9 mm long — and one of the most useful, it is black, with yellowish green abdominal bands, and whitish yellow pollen brush. Growers in the West encourage the bees by providing artificial nesting sites — boxes of malt straws cut in two, sealed at one end with a beeswax-paraffin mixture, and mounted on a platform protected against predators. The bees actually prefer holes in wood, as shown in the picture of a grooved board. Unlike honey bees and bumble bees, *Megachle* bees trip the florets of the alfalfa flower consistently, and thus pollinate them with each visit (99).

The colletid (Collitidae) and andrenid (Andrenidae) bees go to make up the other families of Apoidea. They are solitary bees, usually quite small, and nest in the soil and wood or in clay cells constructed in the crevices of rocks and walls. While numerous and important pollinators of wild plants, they seldom sting anyone.

The Importance of Bees

Bees have great economic significance for man. It is probable that many species of plants we know today would not survive without bees, which in their flower visits unwittingly leave pollen picked up from the stamen of a previously visited flower on the pistil (or female part) of another flower, therby effecting cross-fertilization. Of course, many other insects aid in this process, such as butterflies, syrphid and other flies, wasps and ants; the adults of these insects seek nectar as part of their food. The most important of these pollinators is probably the syrphid or flower flies, which are easily mistaken for bees; being flies, of course, they can't sting, nor are they biters. The flower fly (like other true flies) has only one pair of wings and they are held out half-spread when at rest, instead of folded close to the body as in the case of bees (the latter, of course, have two pairs but this is not always evident at a glance). Another distinction is in the way syrphid flies hover over flowers, while bees weave back and forth and bob up and down.

The importance of honey bees in the production of honey and beeswax has become secondary to their importance in the commercial production of seeds and fruits. Native bees, "because

Fig. 9. The honey bee *(Apis mellifera)* showing pollen basket (Ambassador College photo).

of the destruction of their nesting sites and widespread use of insecticides," are no longer sufficient in numbers to carry out pollination (9). Many **Honey Bee** hives are maintained primarily for cross-fertilization of crops and are moved about and often rented for this purpose. It is estimated that 80 percent of our commercial crops today are pollinated by honey bees. Many forest trees and wild plants, however, are dependent on wild bees (bumble bees, semisocial and solitary bees). Bumble bees are important pollinators of clover in some areas since their long tongues (probosces) can reach through the long corolla of these flowers for nectar; clover flowers are apt to be neglected by short-tongued bees. The particular importance of leafcutting bees and alkali bees in alfalfa seed production has been mentioned.

The Bee Sting

Bees are also important from a medical viewpoint because they "kill more people in this country than any other venomous animal, including rattle snakes." Reports of deaths from stings amount to "probably no more than 25 per year" according to

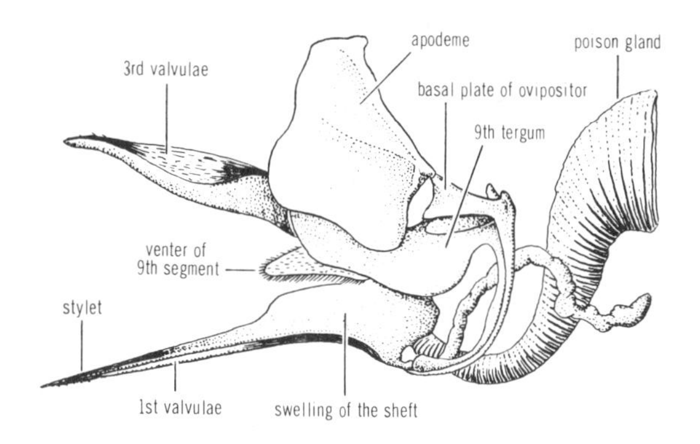

Fig. 10. Anatomy of the honey bee stinger.

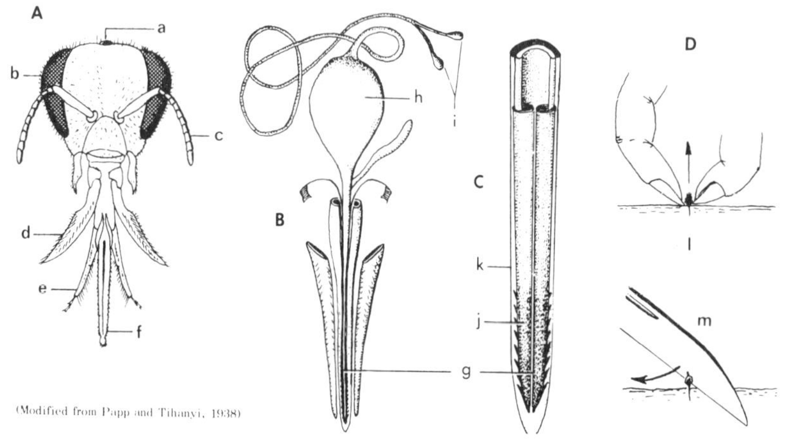

Fig. 11. Mouthparts and stinger of honey bee: A — mouthparts spread apart; B — stinging apparatus (h-poison sac); C — detail of stinger; D — to remove the honey bee's stinger the wrong way (l) and the right way (m).

Frazier, but he thinks that "the true incidence is probably far greater" since the sting is often overlooked and the cause of death reported as heart failure, coronary thrombosis, shock, etc. (31). The **Honey Bee** is the commonest of the stinging insects.

The stinger of the **Honey Bee** consists of a sort of "hypodermic needle" split longitudinally into three parts: a sheath and two lancets, which together form the venom canal. The tip of the "needle," as seen in the detailed drawing of the stinger, has many barbs. The lancets slide back and forth alternately as the stinger penetrates deeper, activated by a bulbous muscular structure at its base and which serves as the plunger. The bulb is connected directly to a small gland, which secretes an alkaline venom, and indirectly to a tabular structure leading to a pair of acid-secreting glands. The two substances are mixed as they enter the stinger canal. The bee's venom is a highly complex mixture of chemical compounds consisting of histamine, several kinds of enzymes and proteins. The protein is thought to be the main cause of the allergic reactions to a sting. The stinging apparatus of other bees and of wasps is essentially the same as that of the **Honey Bee,** without the barbs, and the venoms are similar in nature.

Some persons have a natural hypersensitivity to stings, others develop it after one or more stings, with the sensitivity increasing with each succeeding sting. All beekeepers get stung occasionally but usually "develop some resistance to the ill effects of bee stings" (46). If more than mild irritation and local swelling occur, a physician should be consulted. Symptoms may include a choking sensation and difficulty in breathing, a skin rash similar to hives, a dry cough, sneezing, asthma, lips turning blue, rapid pulse and drop in blood pressure. More severe reaction might include cramps, diarrhea, nausea, vomiting, chills and fever, shock and loss of consciousness. Symptoms usually appear within a few minutes but could be delayed up to 24 hours. Stings about the eyes, nose or throat are the most dangerous.

First Aid and Prevention

As stated before, the stinger of the **Honey Bee** should be removed as soon as possible by scraping with a sharp knife or with the fingernail; it should never be pulled or squeezed since this pumps more vemon into the wound (both bulb and lancets are present, having been torn from the bee's body in its struggle to free itself). [It may be interesting to note that an extract of bee venom was an old remedy for treatment of various ailments, more especially arthritic and rheumatoid conditions, and all kinds of edema involving the accumulation of fluid in the body tissues and cavities. Bee venom was later marketed in ampules, for hypodermic injections, to provide the effect of natural stings without the painful sting.] The site of the sting should be cleaned with soap and water, or by applying an antiseptic.

Squeezing a few leaves of the narrow-leaved plantain *(Plantago lanceolata)* over a sting has a cooling and astringent effect (compresses made from the leaves are an old home remedy for bruises and sprains) (70). Ice packs or cold compresses may help but should never be applied. Infection is more likely to develop in the case of a wasp sting since they are scavengers to some extent. A tourniquet placed between the wound and heart will prevent rapid spread of the venom when the sting is on any of the extremities; it should be released at intervals of three to five minutes and removed when relief comes. When first aid measures of this kind do not bring relief, the help of a physician should be sought.

Immunization by means of long-term treatment with antigens is generally recommended by allergists in the case of highly sensitive persons. Desensitization — as it is called — is said to be helpful in the majority of cases. Since persons known to be sensitive to the venom of one hymenopterous insect are likely to be sensitive to that of the others, a polyvalent extract derived from bees, wasps, hornets and yellowjackets is often used by physicians in the desensitization procedure. Desensitization does not work in the case of ant stings since the reaction is not an allergic one (32).

Bees are affected by the type of clothing people wear. Beekeepers recognize this by wearing light-colored, smooth-textured clothing, which is less likely to irritate the bees than dark or rough clothing. Wooly garments, and especially those of suede or leather, are particularly irritating to bees. Perspiration does not seem to have an effect on honey bees but some perfumes and hair oils are very irritating. Bees will attack a moving object quicker than one at rest, so running, striking or swatting at them is likely to increase the chances of being stung. Beekeepers are careful to avoid jerky movements for this reason. No successful repellent has as yet been developed against bees. Bees are not normally active at temperatures below 55 degrees Fahrenheit, or on rainy days. The greatest number of stings occur during the month of August.

The African Honey Bee

A vicious strain of honey bee not yet present in North America is rapidly spreading northward from Brazil. It was introduced into that country from Africa in 1956 by beekeepers who were attempting to improve the European races then in use. A year later, the bees got out of hand when 26 swarms headed by African queens were inadvertently allowed to escape. This bee — now known as the **Brazilian Honey Bee** because of subsequent hybridization — is dangerous to people and animals since it stings without provocation, and hives are difficult to manage. The workers are usually smaller than those of the European honey bee and their cells are smaller; their flight above flowers is said to be quicker and "more nervous," and they often fly directly into the nest entrance instead of landing on an entrance board and walking in. Despite their smaller size these bees are reported to produce more honey than the Italian race; they start working one-half to two hours earlier in the morning, work later in the evening, and forage at lower temperatures.

"The most alarming and best known characteristic of Brazilian bees is their aggressiveness." They are very sensitive to disturbances and quick to communicate alarm throughout the colony. A hive may "explode" in a matter of seconds, sending hundreds of enraged bees airborne, then they sting and pursue any and all animals and people within a hundred yards of the hive. Disturbed bees pursue a person or animal that has been stung much farther than is usual for European bees. The abundance of these bees — the country-side is said to become filled with small colonies — is cause for alarm. They tend to swarm frequently and rob European colonies (which may account in part at least for their "increased productivity"). A Committee appointed to review the situation concluded that "it is essential to do whatever can be done to minimize the liklihood of this bee moving into North America" (58). If it invades North America it will probably not extend in range beyond the southern states, judging from the experiences in Brazil. The presence of this bee, which is practically indistinguishable from European honey bees by their appearance, will most likely discourage amateur beekeepers from continuing their hobby.

II. WASPS

As in the case of bees, some wasps are social in behavior, but the great majority are solitary; only a few are in between, showing "an approach to social habits." Our social or paper wasps (family Vespidae) number 50+ species and subspecies out of a total of about 600 species and subspecies in the superfamily Vespoidea; members of the still larger superfamily Sphecoidea are all solitary in habit. The social wasps are, however, more conspicuous if not more numerous in places of habitation and account for more than their rightful share of human sting victims. Solitary wasps are those that nest alone, each female constructing cells for the young and provisioning them on her own, without assistance from her daughters or other individual wasps. [An approach to social behavior is shown in the genus *Zethus* in that several females construct a colony of cells made from plant materials glued together, utilizing wood cavities; each female, however, provisions its own cells.] They construct their brood cells in the ground, in hollow stems, in abandoned burrows of other insects and of worms, or fashion them of mud in crevices and on exposed surfaces. These nests are provisioned mostly with caterpillars, the adult forms of various other insects, and spiders. [A few vespid wasps in the genus *Pseudomaris* behave more like bees, in that they provision their mud cells, which are attached to rocks and twigs, with pollen and nectar.] Solitary wasps fall into two main categories: the vespid or potter and mason wasps (family Vespidae), and the sphecids or mud daubers and sand wasps (family Sphecidae). The mud nests of the two groups may be distinguished by the way the egg is placed in the cell; the vespids attach the egg to the wall or ceiling before provisioning is complete or even started, while the sphecids place the egg on the prey after provisioning is started or when it is completed.

Fig. 12. Nests: A — leafcutting bee *(Megachile)*; B — bumble bee *(Bombus)*; C — polistes wasp *(Polistes)*.

The gastronomic preference of sphecid wasps is often quite specific, some species preferring flies, others true bugs, while some concentrate on spiders, others on bees, and some on grasshoppers. Spider wasps, which provide their young with spiders exclusively, comprise a large group known as pompilids (family Pompiliidae, superfamily Vespoidea). The digger wasp (Scolidae) and velvet-ants (Mutillidae) — superfamily Scolioidea — are more closely related to ants. They are parasitic, but the examples given here are known to sting persons. Velvet-ants, it will be noted, are not true ants (Formicidae).

Social wasps — yellowjackets and hornets, polistine *(Polistes)* and polybiine wasps — build paper nests in the ground, in thickets and hollow stumps, in abandoned rodent burrows, or suspended from the limbs of trees and the overhang of buildings. They are similar to the bumble bees in some respects. Like them, they have a well-defined caste system, consisting of a queen and drones or males (the reproductives), and workers (sterile females) that hunt for food and take care of the larvae once a colony has been established by the overwintering queen. Workers can lay eggs, and sometimes do toward the end of the season, but they are unfertilized and produce males only. After the first brood emerges as adults, the queen confines her efforts to egg laying. As with her counterpart among the bees, she has the ability to regulate the sex of her progeny — fertilizing the eggs intended to become females, and not fertilizing those which her cerebral blueprint tells her will be required to perform the male function. As in the case of bumble bees, the colony breaks up in the fall and only the young mated queens survive and hibernate.

Paper Nests

The paper nests of the social wasps — made from wood chewed to a pulp and mixed with saliva — resemble those of the honey bee in construction, being composed of rows of hexagonal cells in which the eggs are laid and the young develop. The cells of hornets' or yellowjackets' nests are generally enclosed in a paper covering, with only a hole provided for entrance and exit; those of the polistine and polybiine wasps are exposed and the heads of the larvae plainly visible until they pupate. Their most conspicuous departure from the bees is in the food provided for

Fig. 13. A polistes wasp on brood comb. Ends of cells have been sealed over by the larvae, preparatory to pupating (USDA photo).

the young. The queen and later the workers prey on other insects, prepare and feed them to the developing larvae continuously. The white legless larvae are fed partially masticated insects, consisting of caterpillar and adult forms, the latter often being caught on the wing. From an economic viewpoint the wasps are beneficial since their forays to feed the larvae sometimes reduce the pest population significantly. The adult wasps themselves feed on the nectar of flowers, on ripe fruit, or bits of the insects captured for the young. They are also fond of the secretions coming from the mouths of the larvae, and often withhold their food until this delicacy is offered. They have well developed mandibles, suited to their predatory habit, and tongue for sucking the nectar from flowers and the juices from soft fruits.

Yellowjackets and Hornets

There is a great deal of confusion in the literature as to what is a "yellowjacket" and what is a "hornet." According to Duncan, the only true hornet in North America is the **Giant Hornet** *(Vespa crabro germana),* which was introduced from Europe and established only in the eastern part of the country. "The true hornets have the head swollen behind the eyes and the ocelli remote from the margin of the head" (25). The rest of our species are accordingly "yellowjackets," though some are black and white. [Yellowjackets have been separated by some writers into two genera, *Vespula* and *Dolichovespula,* which are now considered subgenera of one genus, *Vespula.* The distinction is a technical one: in the subgenus *Vespula* the oculo-malar space (between the lower margin of the compound eye and the base of the mandibles) is less than half the length of the penultimate (second from last) antennal segment, in *Dolichovespula* it is more than half. Stating it more simply, if less accurately, *Vespula* has a short face, *Dolichovespula* has a long face.] The **Giant Hornet** is a large brown and yellow species known as the brown hornet in Europe, from 17 to 23 mm long, usually nests in hollow trees or stumps. One of the most common and widely distributed wasps in North America is the so-called **Baldfaced Hornet** *(Vespula maculata),* a black species with white areas on the face and front corners of the thorax, and white markings on the tip of the abdomen as shown; it is from 12 to 18 mm long, builds a large globular nest suspended from the branch of a tree or bush, building overhang, sometimes the ceiling of a garage or barn. The closely related *V. arenaria* is a common and widespread yellowjacket, occurs across the southern part of Canada and most of the northern states, and throughout the western states; it is black, with yellow markings as shown, from 13 to 17 mm long, builds aerial nests, often under the eaves of houses where it becomes a pest. Both of these species are in the subgenus *Dolichovespula,* have a long oculo-malar space.

These wasps are considered vicious when distrubed but an early settler, Hector St. Jean de Crevecoeur, tells of their use to rid his house of flies (this was before screens came into use). He hung a limb with attached nest in the middle of his parlor, leaving an exit hole in one of the window panes for the wasps to

Fig. 14. Paper-nest wasps: A — *Vespula consobrina;* B — *Vespa crabro germana,* the feared giant hornet; C — *Vespula maculata.* The bald-faced hornet; D — *Vespula pennsylvanica,* the well known yellowjacket.

come and go as they pleased. The family and wasps got along very well together; the children allowed them to take flies off their eye lashes without flinching, confident that their benefactors would not betray their trust (23).

The **Pennsylvania Yellowjacket** *(Vespula pennsylvanica)* is common in the western part of the continent, black, with yellow markings as shown, upper three-fourths of compound eyes (which are set close to the mandibles) encircled with a yellowish band, from 12 to 17 mm long; its nest is terrestrial. The **Common Yellowjacket** *(Vespula vulgaris)* is widely distributed across the continent, except the southern states, black, with yellow markings, from 12 to 18 mm long; it is one of the commonest species, its nest terrestrial. *V. consobrina* ranges across Canada and most of the northern states, including the Northwest, has also been reported from Redlands, California; it is black, with white markings, uniformly marked throughout its range, its nest terrestrial as a rule but sometimes just above the ground in bushes. *V. sulphurea* is almost completely restricted to the lower foothill areas of California, also occurs in Baja California, and is reported from Arizona, Nevada and Oregon; it is black, with numerous yellow markings, from 12 to 17 mm long, nests in the ground. The above four species are in the subgenus *Vespula,* have a short oculo-malar space.

Paper Wasps

The paper or polistine wasps (family Vespidae) resemble yellowjackets in color patterns, but they have a more slender body and a short pedicel or waist connecting the abdomen and thorax. Another distinction is in the small lobe on the outer margin at the base of the hindwing in the paper wasps, which is absent in the yellow jackets. Their nests are usually much smaller than those of the yellowjackets but make up in number for their size, if sufficient nesting sites are available; when encouraged they have been helpful in controlling caterpillars. The nest consists of a circular comb of hexagonal paper cells suspended from an overhang by a short stem. The ends of the cells are open and the heads of the larvae plainly visible during the feeding process; the cells are closed by the larvae when they are ready to pupate. The closely related polybiine wasps are easily separated by the long petiolate basal segment of the abdomen and the gaster, which resembles that of ants. The nests are similar to those of polistine wasps but somewhat smaller, and are similarly attached to natural or man-made supports.

The **Zebra Paper Wasp** *(Polistes exclamans exclamans)* is widely distributed from Florida to Maryland, west to Iowa, Nebraska, Colorado, California and Mexico; it is reddish brown, abdomen with blackish, reddish and yellowish markings, mesonotum reddish, head with yellowish band behind the ocelli, about 15 mm long, builds single or double paper combs under eaves or natural overhangs. This species is sometimes seen in great swarms around trees in the fall. *P. fuscatus aurifer* ranges from California to Idaho and British Columbia and Arizona; it is black and yellow, from 9 to 15 mm long, hibernates in the company of *P. hunteri californicus* under the loose bark of cottonwood and eucalyptus trees. *P. hunteri californicus* occurs in California, Arizona and Baja California, is pale reddish, with yellow markings, from 10 to 14 mm long; its nests are usually small but may have as many as 250 cells and double pedicel. The polybiine wasp *Mischocyttarus flavitarsis flavitarsis* is yellowish orange, reddish, and black, from 16 to 20 mm long, builds a single free comb of hexagonal paper cells similar to that of *Polistes;* it occurs in California, Oregon, Utah, Colorado, Nebraska and Arizona.

Potter and Mason Wasps

The potter wasps (family Vespidae) are usually black, with yellow markings, and medium to large in size. They make mud nests resembling a small jug on twigs of trees and bushes. Several jugs may be constructed along a single branch. The jug is filled with paralyzed caterpillars (after an egg is placed on the wall or ceiling) and is then sealed. The potter *Eumenes crucifera* occurs in western United States and Canada; it is black, body covered with fine golden pile, tibiae yellow, yellow marginal bands on abdominal segments, yellow spot on each side of petiole and second segment of gaster, wings brownish transparent, about 14 mm long. The subspecies *E. crucifera nearcticus* ranges from Canada to Oregon, Utah and Colorado, *E. crucifera flavitinctus* is found in California. *E. bollii* is widespread throughout the western half of the United States; it is black, with dense yellowish pubescence, legs yellowish brown, tibiae and tarsi yellow, abdominal segments banded with yellow, from 12 to 14 mm long. The subspecies *E. bollii oregonensis* occurs in Washington, Oregon, Idaho and Nevada. The species in this genus are characterized by the petiolate abdomen.

The mason wasps are also black and yellow as a rule, from medium to large in size; they nest mostly in wood and other cavities, partitioning off a series of cells using mud. They take over abandoned tunnels of carpenter bees and ground-nesting bees, and old nests of mud daubers. The mason *Monobia quadriens* is a large black and yellow species, about 20 mm long, provisions its cells with large cutworms; it occurs in the eastern and southern states, west to New Mexico. *M. texana* is reddish brown, with short silky, golden pile, long abdomen, the first segment black at base and dull yellowish at apex, wings dark brown, about 16 mm long; it is found in Texas, New Mexico and Arizona.

In the closely related genus *Odynerus,* nests are usually made in the ground; *O. erythrogaster,* a species found in the foothills of California, is an exception, and utilizes the twigs of elderberry; it is black, with first five abdominal segments bright orange-red, the head, thorax, and first abdominal segment covered with long black hairs, from 9 to 11 mm long. *O. cinnabarinus* is very similar to the above species in coloring, but hairs on head, thorax and abdomen are sparse; it is from 10 to 11 mm long, occurs in California, Utah, Arizona and Texas. The masarine wasp *Pseudomasaris vespoides,* a beautiful black and yellow mason wasp, from 15 to 22 mm long, behaves more like a bee than a wasp. It fashions a series of cylindrical mud cells plastered upright to rocks, anywhere from 2 to 13 cells in a row, the tops even, insides lined with silk and stored with pollen; it occurs in the western states, also South Dakota and Mexico.

Spider Wasps

Spider wasps (family Pompilidae) are usually black, but many have metallic blue, orange or red coloring. They are small (most of them) to large in size and very active, with long legs enabling them to run rapidly on the ground or over plants. The **Tarantula Hawk** *(Pepsis mildei)* is a large species, from 20 to 30 mm long. It is metallic blue-black with reddish antennae and fiery red wings (excepting the dusky bases and apexes), occurs from Texas to Kansas, west to California, where it preys on the so-called tarantulas (family Theraphosidae). Each species of *Pepsis* attacks only a certain species of tarantula. The female wasp approaches the much larger tarantula slowly, while the spider makes no effort to resist or escape, and even tolerates a careful and rather prolonged exploration of its body by the wasp's antennae, which serves to make the necessary identification. If it's the right one, she prepares a burrow nearby (meanwhile keeping an eye on the spider), then returns and seizes the spider's leg with her powerful jaws; after a desperate struggle, she invariably manages to insert her stinger into the soft tissue at the junction of the spider's leg and body. [The sequence varies with species and individuals within a species. Some *Pesis* wasps paralyze the spider before excavating the burrow. Sometimes the tarantula is flushed from its burrow, which may

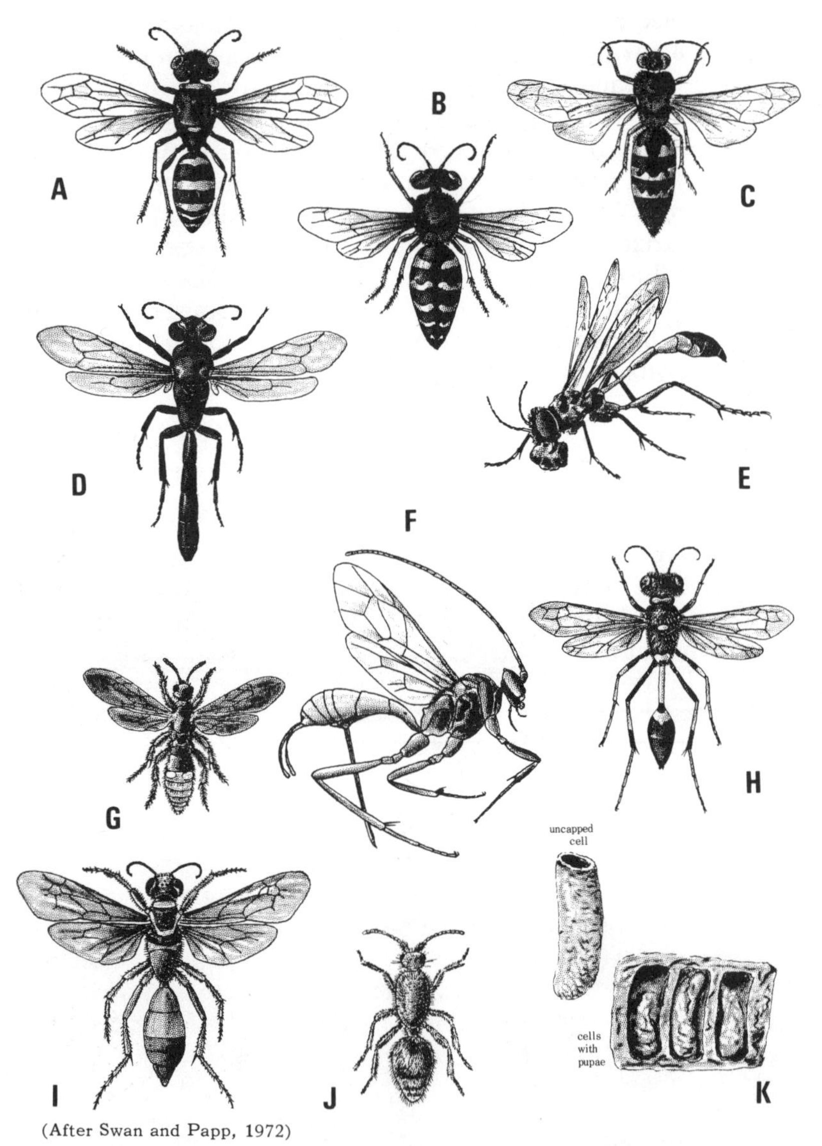

(After Swan and Papp, 1972)

Fig. 15. Some solitary wasps: A — a burrowing wasp *(Philanthus ventilabris);* B — Eastern sand wasp *Bembix spinolae);* C — giant cicada killer *(Sphecius speciosus);* D — mud dauber *(Trypoxylon clavatum);* E — sand wasp *(Sphex urnarius)* tapping dirt with a stone; F — an ichneumon wasp *(Cryptanura branchiformis);* G — digger wasp *(Scolia dubia);* H — black-and-yellow mud dauber *(Sceliphron caementarius);* I — great golden digger wasp *(Chlorion icheumonea);* J — two-spotted velvet ant or mutillid wasp *(Dasymutilla bioculata);* K — mud cells of black-and-yellow mud dauber (H).

then be used by the spider instead of excavating one of its own (16).] The paralyzed spider is then dragged across the ground and over obstacles and eventually down to the cell at the end of the burrow, where an egg is placed on it; the burrow is now painstakingly filled by the wasp with bits of dirt carried in its jaws (69). *P. formosa formosa* is one of the largest of these wasps, from 20 to 40 mm long, occurs from Texas to Kansas, west to Arizona and Nevada; it is metallic blue and black, including the antennae. *P. attoni* occurs in New Mexico, Arizona and California; it is a large black species, sometimes 42 mm long, with black wings having narrow pale band at tip, prominent side tubercles on first abdominal segment.

Mud Daubers

Mud daubers (family Sphecidae) make nests of mud in sheltered places and provision them with various kinds of spiders. The **Common Mud Dauber** *(Sceliphron caementarium)* is a very large black and yellow, thread-waisted wasp, from 30 to 35 mm long, often seen around mud puddles forming balls to be used in the construction of its nest; one of the most common and widespread of the mud daubers, it ranges across southern Canada and the entire United States. Its nests are usually

Fig. 16. Some mud dauber wasps: A — pipe organ mud dauber *(Trypoxylon politum)* and mud cells; B — blue mud dauber *(Chalybion californicum)*.

composed of several cells formed on the sides of rocks or logs, on fences or garages; after filling a cell with paralyzed spiders, packing them in with her head, the female lays an egg on the food mass and seals the cell. The **Blue Mud Dauber** *(Chalybion californicum),* equally common and widespread, is metallic blue and blackish, with blue wings, from 12 to 18 mm long. It lacks the skill or inclination to make its own mud nest, and appropriates one newly built by the **Common Mud Dauber;** it moistens the clay with water to open up the cell, and throws out the stores and egg placed by the other wasp. With this it proceeds to deposit paralyzed spiders of its own, placing an egg on the first one, then filling and sealing the cell. The **Blue Mud Dauber** prey mostly on black widow spiders and is probably their worst enemy.

Sand Wasps

Sand wasps (family Sphecidae) build their nests in the soil and are often quite specific in their choice of insects provided as food for the larvae. The **Western Sand Wasp** *(Bembix comata)* ranges from British Columbia to California, Alberta, Nevada and New Mexico, is common on the sand dunes along the seashore; it is black, with long dense pubescence, interrupted abdominal bands yellowish in the female, white in the male, from 10 to 17 mm long. It digs an inclined burrow about 10″ in depth, with a single brood cell at the end; the larva is provided with flies "progressively," that is, at intervals during the developmental period, the burrow being opened and closed with each entry and exit of the provident female. A branch tunnel with cell at the end is dug from the existing main burrow for each succeeding brood. The **Great Golden Digger** *(Chlorion ichneumonea)* is a beautiful black and gold burrowing wasp, about 25 mm long, widespread across southern Canada and the entire United States. It digs a burrow about 12″ in depth, with several cells radiating from a common chamber; each cell is provided with an average of three grasshoppers and an egg is placed on the first one in each case. Another common and widespread burrowing wasp is the **Bee Wolf** *(Philanthus ventilabris),* generally distributed throughout Canada and the United States; it is black, with yellowish abdominal bands, yellow legs and wings with yellowish cast, about 12 mm long. Its nests are oblique burrows with several branches, each ending in a brood cell; the cells are provisioned with several kinds of bees. This large genus as a whole constitutes the principal predaceous enemy of bees.

Parasitic Wasps

The scoliid wasps (family Scoliidae) are external parasites of the larvae of scarabaeid beetles. They are fairly large wasps with stout legs, and have been known to sting people. The **Digger Wasp** *(Scolia dubia)* is widely distributed from Florida to Massachusetts, west to Colorado and Arizona; it is black, hairy, with bright red and yellow markings on the abdomen, and black wings with membranes beyond the cells wrinkled, from 12 to 18 mm long. Males and females perform a mating dance, flying close to the ground on a horizontal plane and in a figure 8 course. The female burrows into the soil in search of grubs of June beetles, often stinging more of them than she can parasitize (the attack in any case is fatal to the grub). When a suitable host is found, the wasp burrows deeper — sometimes as much as four feet — dragging the grub after her; at the bottom she forms a cell and leaves the grub implanted with an egg. The subspecies *S. dubia haematodes* is found in Texas, New Mexico, Arizona and California. *S. dubia monticola* occurs in Mexico, north to western Texas, New Mexico and Arizona.

The velvet-ants (family Mutillidae) comprise a fairly large group of external parasites that attack the larvae or pupae of bees, vespid and sphecid wasps, mostly the ground-nesting forms. Males are usually winged and larger than the females. The latter are wingless and resemble ants; they may be distinguished from ants by the absence of the dorsal node or projection on the petiole which characterizes ants, and by the dense coating of brightly colored hairs. The females use their stingers for defense against adult bees and wasps when invading their nests. They can run rapidly and may sting people with painful results if molested. Some of them emit a plainly audible squeak when disturbed. **Sacken's Velvet-Ant** *(Dasymutilla sackenii)* is found from Oregon to California, Arizona and Nevada; it is black, with dense coat of white hairs, about 12 mm long. The **Gray Velvet-Ant** or **Thistle Down Mutillid** *(D. gloriosa)* is similar to *sackenii,* black, with long white hairs, from 12 to 16 mm long; it occurs in Texas, Arizona, Nevada, Utah and California. *D. bioculata* ranges from Manitoba south to Louisiana, west to British Columbia and New Mexico; it is black, with dorsum of thorax and abdomen reddish, from 17 to 30 mm long. This species varies greatly in size, according to the size of the hosts (sand wasps); those feeding on larger hosts become proportionately larger.

An Ounce of Prevention

Being "loners" usually, and not so numerous or evident in places of habitation, solitary wasps are in general much less troublesome than honey bees and social wasps. Many of them nevertheless get in our way (or is it the other way around?) and a few of them are quite large and capable of inflicting a painful sting. The females of many velvet-ants have a long curved stinger and a potent venom, which they readily inject into humans if handled or stepped on with bare feet. They are frequently found on beaches where bathers are sometimes stung by them. The most formidable of the solitary wasps is the **Tarantula Hawk.** In southern California these wasps are quite common around houses in suburban areas. One sees a female not infrequently crawling up a wall or steep embankment dragging its huge prey with seeming ease. They are more commonly seen soaring a few feet above the ground searching for prey; at such times one should try to keep a safe distance away. When ready to attack, the female *Pepsis* gives off a pungent odor. Its sting is said to be "much worse than that of a bee or common wasp, and the pain and swelling last longer" (69).

Mud daubers are common around houses and often build their mud nests on fences and walls, under overhangs, and even on garage doors with suitable corners or niches. Unlike the paper nests, the mud nests do not disintegrate and the wasps overwinter in them as pupae. We used to have large pipe organ nests on a clapboard siding (under an overhang) near the front entrance to the house; while much of the family outdoor activity centered there. This seemed to make no difference to the wasps. If you can't live with them, an insecticide can be applied when the wasp makes one of her visits with a captive spider, or at night. Another way to get rid of them is to remove the nest, first stunning the wasp by plugging the entrance with a wad of cotton treated with a few drops of carbon tetrachloride or formaldehyde.

Wasp Stings

The venom of wasps and its reaction are similar to that of the social bees (see I. Bees), though it appears that the reaction is generally more severe and more apt to produce shock in the case of yellowjackets. Persons known to be sensitive to bee venom are likely to be sensitive to wasp venom, apparently because of the common antigens (31). The stinging apparatus of wasps is essentially like that of the honey bee, as can be seen by comparison of the drawing showing the anatomy of the honey bee's stinger (Fig. 10) and that of the *Polistes* wasp (Fig. 17). The main points of difference are: the wasp's stinger is larger, is slightly barbed on one side only and usually withdrawn without difficulty, and is activated by muscles around the venom sac rather than by a plunger.

Yellowjackets inflict more stings than bees or hornets and other wasps (33). Their nests are often concealed by bushes, from which they attack the passerby without warning. We had such a nest in our backyard and not one of the family was bothered by the wasps. A young neighbor however was immediately set upon by the yellowjackets when he came near, and again sometime later, having failed to recall soon enough his first painful experience. We realized then the necessity of removing the nest, and the possibility of varying degrees of "attractiveness" (or irritation) of different persons to the wasps.

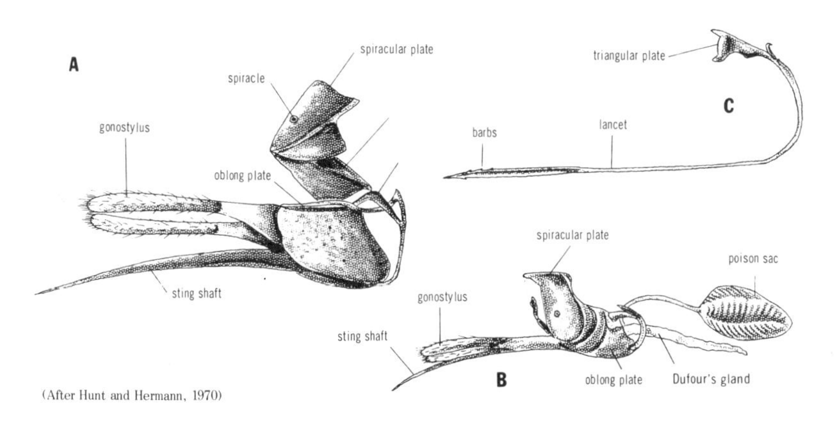

Fig. 17. Anatomy of the *Polistes* wasp's stinger.

They frequently intrude on picnickers when soft fruits and juices are left exposed. The situation is different than when one threatens their nests, and they will normally leave peacefully if the attraction is removed and they are not irritated. As with bees, jerky movements and swatting at them will invite stings. Though generally more peaceful and less vicious than yellow-jackets, *Polistes* wasps are sometimes troublesome to citrus workers and often sting them. Their nests are easily spotted under the eaves and overhangs of houses and other buildings, less so under limbs of trees.

The first aid measures suggested for wasp stings are the same in general as for bee stings (see page 24). Besides ice packs, a USDA leaflet suggests application of "a paste made of water and bicarbonate of soda" as a first aid measure for wasp stings (93). As stated before, the stings of yellowjackets and hornets are more likely to cause infection than those of bees since wasps are scavengers to some extent.

Removal of Paper Nests

Preventive measures would of course include removing nests when they are close to or around houses, particularly where sensitive persons reside or are apt to come in contact with the wasps. However, one should consider, in our view, that these are valuable predators of insect pests and avoid destroying their nests unnecessarily, more especially those of *Polistes* wasps. It may be necessary to use an insecticide to destroy a colony. Dusts are easiest to apply, using a simple hand duster, whether the nests are above or below the ground. Two or three strong puffs of dust will filter through the nest and usually destroy the inhabitants within 24 hours. A shovelful of moist dirt thrown over the entrance to an underground nest after dusting will prevent escape of the wasps. Several ounces of carbon tetrachloride or formaldehyde poured into the entrance to an underground nest, and plugging it with cotton, will also kill the wasps. One should be very careful not to inhale the insecticide dust or the fumes of the carbon tetrachloride; or a protective mask should be worn. The work should be carried out at night, when foraging activities have come to a halt and the wasps are in their nests.

Paper nests can sometimes be removed in a sack or other container, first plugging the nest opening with cotton soaked in carbon tetrachloride or formaldehyde to stun the wasps. If you are strongly oriented ecologically, and protect yourself, you can take the nest to a remote area and release it without further harm to the wasps rather than destroy the colony. Gaul moved the nest of a yellowjacket *(Vespula arenaria)* a mile away, after first anesthetizing the colony. When the adults recovered, they abandoned the nest with its young and returned to the old site and started all over again (34). (The second go at it results in an inferior nest, as if in the interval the wasps had forgotten something or lost some of their skill.) When kept in a box, or in some way not allowed to escape for a day at the new location, they continued as before with the rearing of their young, as if nothing had happened. In the case of an aerial nest, one might emulate Hector de Crevecoeur and remove the limb and nest (see page 33). As mentioned before, all colonies of social wasps break up in the fall and only the mated queens survive to found a new colony the following season. The old paper nests are never reused, and usually disintegrate during the winter or are torn apart by birds and squirrels searching for food or nesting materials.

III. ANTS

Ants (family Formicidae) are found practically everywhere and are one of the most abundant insects. There are over 700 species and subspecies known in North America north of Mexico — not a particularly large group as insect families go, but certainly one of the most successful judged by the number of individuals and their prevalence. They are black, brown, red or yellow or some shade of these colors, and are characterized by the petiole or waist connecting a greatly enlarged portion of the abdomen (called the gaster) and the thorax, and by the large head and elbowed antennae. The petiole constitutes the base of the abdomen and may consist of one or two segments, according to the kind of ant it is; these segments have a characteristic node or projection on the top.

Ants are sometimes confused with termites (which unfortunately are sometimes called "white ants"), but they are easily distinguished. The body segments of the termites are more or less uniform in width and there is no petiole. In the winged forms, termites are readily separated by the length of the fore- and hindwings, which are *equal* and extend far beyond the end of the body when folded (89); in the ants they are *unequal* in length and barely extend beyond the end of the body when folded. Ants nest mostly in the ground, but some "house ants" nest between the walls of buildings. Termites live in wood, which they actually eat, though the subterranean ones (in contrast to the dry wood termites) must also have a connection with the soil. Carpenter ants also nest in wood, which they do not eat, and maintain contact with the ground.

Ants are social, and have three basic castes, as social bees and wasps do: the workers or infertile females, the queen, and drones or males. The queens and drones — the reproductives — have two pairs of wings and gather together in huge swarms to mate, which they do in flight or on the ground. After mating the male dies and the female gnaws off her wings, digs into the ground to start a new nest and to begin egg laying. Workers are monomorphic, dimorphic or polymorphic having one, two or more forms. Worker ants can also lay eggs and some are believed to be capable of producing females even though unfertilized. The workers, being the most numerous of the castes, and the foragers, are the ones commonly seen. They are also easier to identify and the descriptions and illustrations pertain to them unless otherwise indicated (84).

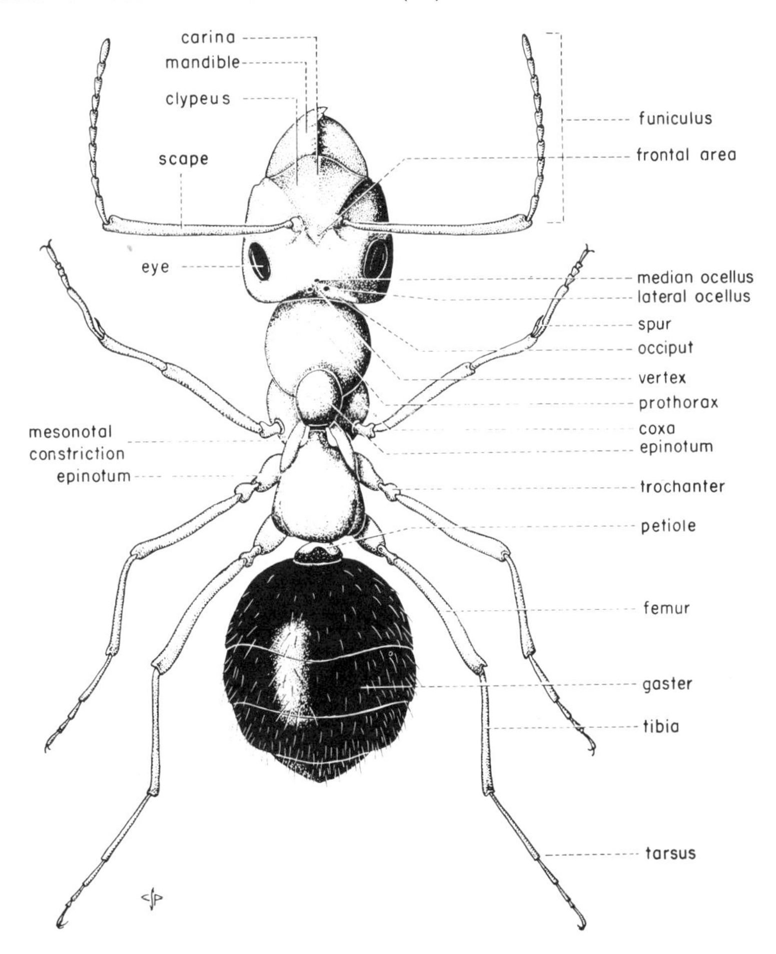

Fig. 18. Detail of an ant. Top view, showing large head with toothed mandibles and peculiar antennae (note long scape and many-segmented funiculus), elongated and distinctly-segmented thorax, petiole or waist joining the thorax and enlarged abdomen or gaster, leg segments and other features.

The majority of ants are herbivorous but many are carnivorous and predaceous on injurious insects. Ants in general are more of a nuisance from their intrusions into homes, where they go in search of sweets and meats, than from their stings. The fire ants and harvester ants cause most of the trouble (and sometimes plenty) so far as the stingers are concerned. Some ants can sting and bite, others will only bite. According to Wheeler, the stinger is well developed in the subfamilies Dorylinae (legionary or army ants), Ponerinae (a small group of primitive ants), and in most of the Myrmicinae (a large diverse group containing most of the pest species); in the Dolichoderinae (a small group mostly southern in distribution), and Formicinae (the second largest group), it is vestigal or absent (100). Ants are not believed to be implicated in the transmission of disease organisms affecting man. [A "bigheaded ant" *(Pheidole bicarinata vinelandica)* — a myrmicine ant found in the central states, which feeds on seeds and small insects outdoors and meats, grease and bread indoors — is an intermediate host of a tapeworm of wild and domestic fowl.] The **Argentine Ant** *(Iridomyrmex humilis)* — the workers have no sting and only a feeble bite — may be an exception, since it goes everywhere once it gets into a dwelling, from bathrooms and toilets to kitchens. The workers of this species are monomorphic, brownish in color, with one-segmented pedicel, from 2.2 to 2.6 mm long; they occur in the southern states and California, where they are agricultural pests as well as household pests, since they are very fond of the honeydew of aphids, scales and mealybugs and protect them from their natural enemies.

Legionary or Army Ants

The legionary ants occur mainly in the southern and southwestern states and southward. They differ from other ants notably in that the winged males are wasplike in appearance, the queens are always wingless, with petioles one-segmented instead of two-segmented as in the workers, and the workers are often blind. Colonies are normally very large, nests are temporary and nothing more than a cluster of ants, usually in a decayed stump or under a log. They are periodically nomadic and in fixed bivouac, in rhythm with the queen's reproductive cycle. Stationary periods are for egg laying and emergence of the callow workers, during which the workers conduct daily raids for food. Migrations take place at night; in migrating columns, larvae and pupae are carried along, in raiding columns they are not. They are ferocious predators and devour all insects and small animals in their path, and sometimes kill chickens and small pets that get in the way. The workers are polymorphic and can bite and sting viciously.

Our most common and widely distributed legionary ant is *Neivamyrmex nigrescens.* [The genus is known in the older literature as *Eciton.*] It ranges from Virginia to Florida, west to California, north to Iowa, Nebraska and Colorado, is light brown to reddish brown and almost black, body excepting gaster opaque, covered with dense granular punctures interspersed with coarse pits, from 2.8 to 5.8 mm long. The closely related *N. opacithorax* ranges from Kansas to Virginia, south to Florida, Arizona, California and Baja California; it is similar

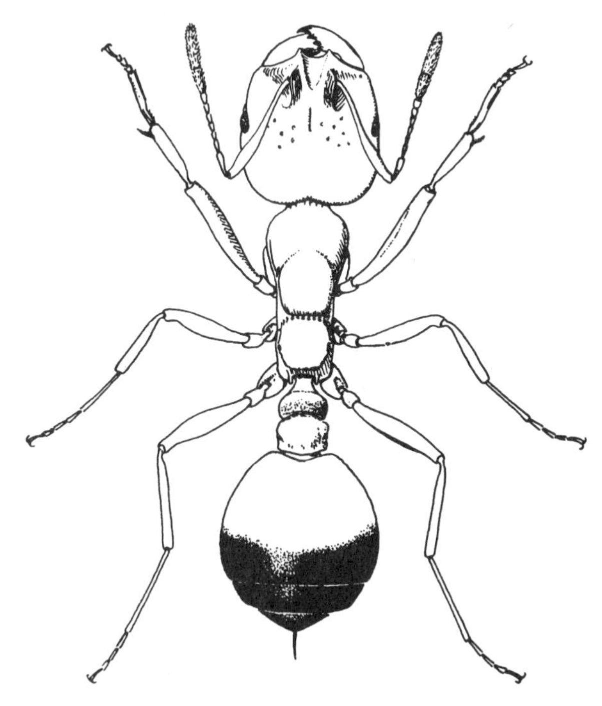

Fig. 19. Top view of the fire ant *(Solenopsis geminata)* showing chewing type mandibles and stinger (USDA).

to *nigrescens* except that the antennae are not so stout, scape not quite as long, eyes less distinct, and only the thorax and petiole are opaque and densely sculptured; it is 2.2 to 4.6 mm long. Many of the foraging activities of these species take place in daylight. They are important predators of the **Argentine Ant,** and sometimes enter houses; the males leave their nests from September to November. *N. californicus* is common around Sacramento and the San Francisco Bay area in California; the workers are active at night, gather around lights where they prey on other insects attracted to the lights.

Harvester Ants and Others

The harvester ants (genus *Pogonomyrmex)* are found in the western states mostly, and feed on seeds and grains. They have a two-segmented pedicel, and are generally thought to be monomorphic but there appears to be some division of labor at least in some species. The **Western Harvester Ant** *(Pogonomyrmex occidentalis)* is a large reddish brown species about 6 mm long; it occurs from British Columbia and Washington to North Dakota, south to Oklahoma and Arizona. The large nest is in time marked by a huge mound resembling an inverted cone, four or five feet in diameter at the base and two or three feet high; there is usually only one entrance, facing the east, sometimes a second one facing the south (50). The ants clear a space around the mound by clipping the grass, denuding an area from eight to 30 feet in diameter (the latter where the vegetation is sparse). This sometimes results in a loss of one-seventh of the vegetation on rangeland and exposes it to severe wind erosion. Swarming occurs mostly in July and August but may be anytime from April to October.

The **Red** or **Texas Harvester Ant** *(P. barbatus)* is reddish brown, and the largest of the harvester ants, from 6 to 12 mm long. The subspecies *P. barbatus rugosus* occurs in California, Arizona, New Mexico, Utah, Texas, Colorado and Kansas. The nest is marked by a bare circular disc with the entrance in the center, rather than by a mound as in *occidentalis;* the disc is from three to ten feet in diameter and formed by cutting down and removing the vegetation. The **California Harvester Ant** *(P. californicus)* occurs from Texas to California and Nevada; it is reddish brown, from 5 to 7.5 mm long. The nest is built in sand and fine gravel, with a fan-shaped crater on one side of the entrance; no vegetation is cut down as in the case of the other two species. Like the legionary ants, the harvester ants bite and sting viciously. The sting is painful, usually causing severe swelling and irritation, or worse in hypersensitive persons.

The **Black Harvester Ant** *(Veromessor andrei)* occurs in California, Nevada, Arizona and Mexico; it is black, with brownish thorax and pedicel, latter two-segmented, long epinotal spines, from 5 to 6 mm long. It forms large colonies; the nests usually have a single crater (rarely more), about two feet in diameter, flattened on top, with rounded slopes and one to three large irregular openings. Long files of workers may be seen in the afternoon carrying mature seeds of all kinds back to the nest where they are husked and stored in the granaries; the chaff and discarded seeds are carried out and deposited on the periphery of the crater, where they form a circle and eventually result in a fringe of grass. *V. pergandei* is similar, shiny brown to black, clypeus with prominent projection in middle of anterior border, about same size as *andrei* and with similar habits, forages in long files. It is strongly polymorphic, very common in the deserts of southern Arizona and Nevada and the Mojave Desert of California.

The **Pyramid Ant** *(Conomyrma insana)* [This species, previously known as *Dorymyrmex pyramicus,* was the only one found by Wheeler on the walls of the Grand Canyon "from the rim to the river," a succession of temperature zones "equivalent to those stretching from the coniferous forests of northern Canada to the cactus plains of Mexico."] is a monomorphic species, light brown to blackish (gaster often darker than remainder of body), with dense pubescence on thorax and gaster, and one-segmented pedicel as in other dolichoderine ants, about 3 mm long; it ranges from Oregon to New York, south to California and Florida. Colonies are small to medium in size and frequently established near those of the harvester ants *Pogonomyrmex occidentalis* and *barbatus.* The fast-moving workers forage in long conspicuous columns; they are carnivorous but fond of honeydew, and become pests by building unsightly nests on lawns and invading houses. There have been reports of children being bitten by them.

The **Western Thief Ant** *(Solenopsis molesta validiuscula),* a close relative of the fire ants, occurs from California to New Mexico, Utah, Idaho and Washington; it is yellowish, with two-

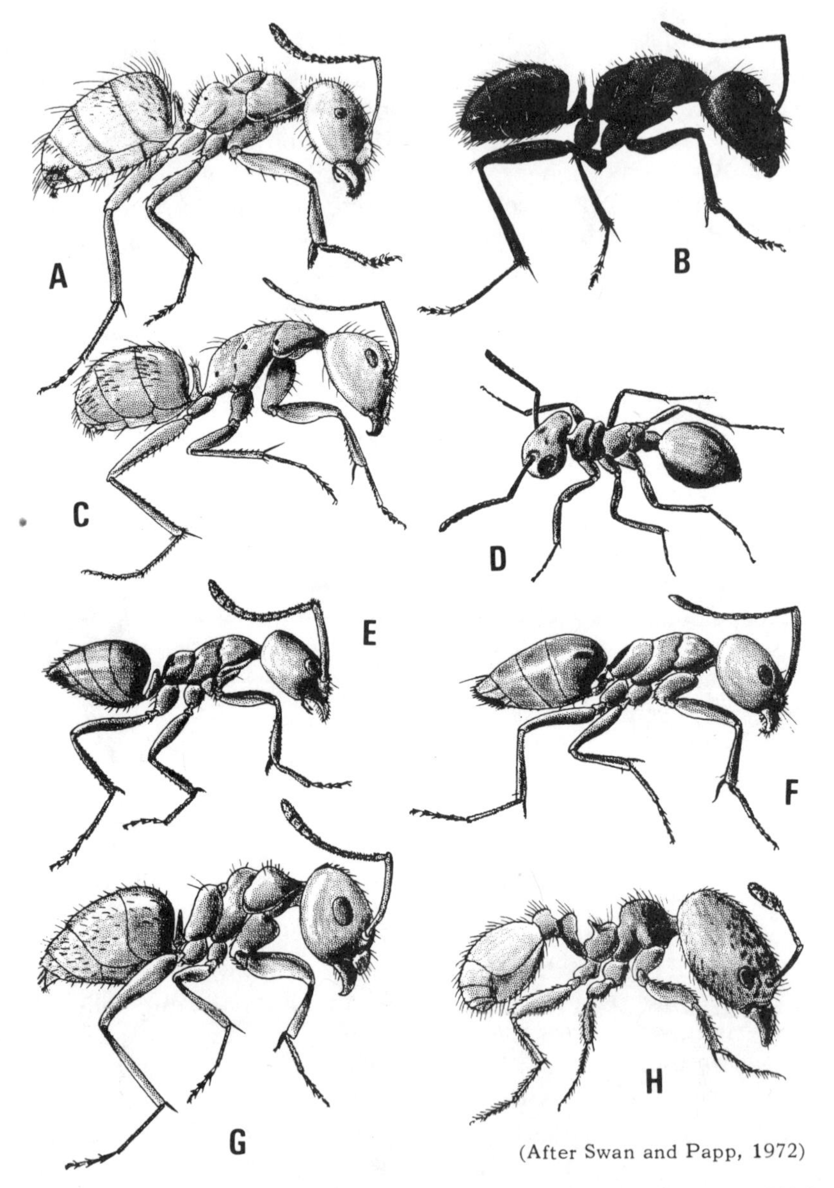

(After Swan and Papp, 1972)

Fig. 20. Ants: A — larger yellow ant *(Acanthomyops interjectus);* B — black carpenter ant *(Camponotus pennsylvanicus);* C — brown carpenter ant *(Camponotus castaneus);* D — legionary ant *(Labidus coecus);* E — Argentine ant *(Iridomyrmex humilis);* F — odorous house ant *(Tapinota sessile);* G — cornfield ant *(Lasius alienus);* H — big-headed ant *(Pheidole bicarinata vinelandica).* It will be noted that all of these ants excepting D and H, have a one-segmented petiole or waist; D and H have two-segmented petiole.

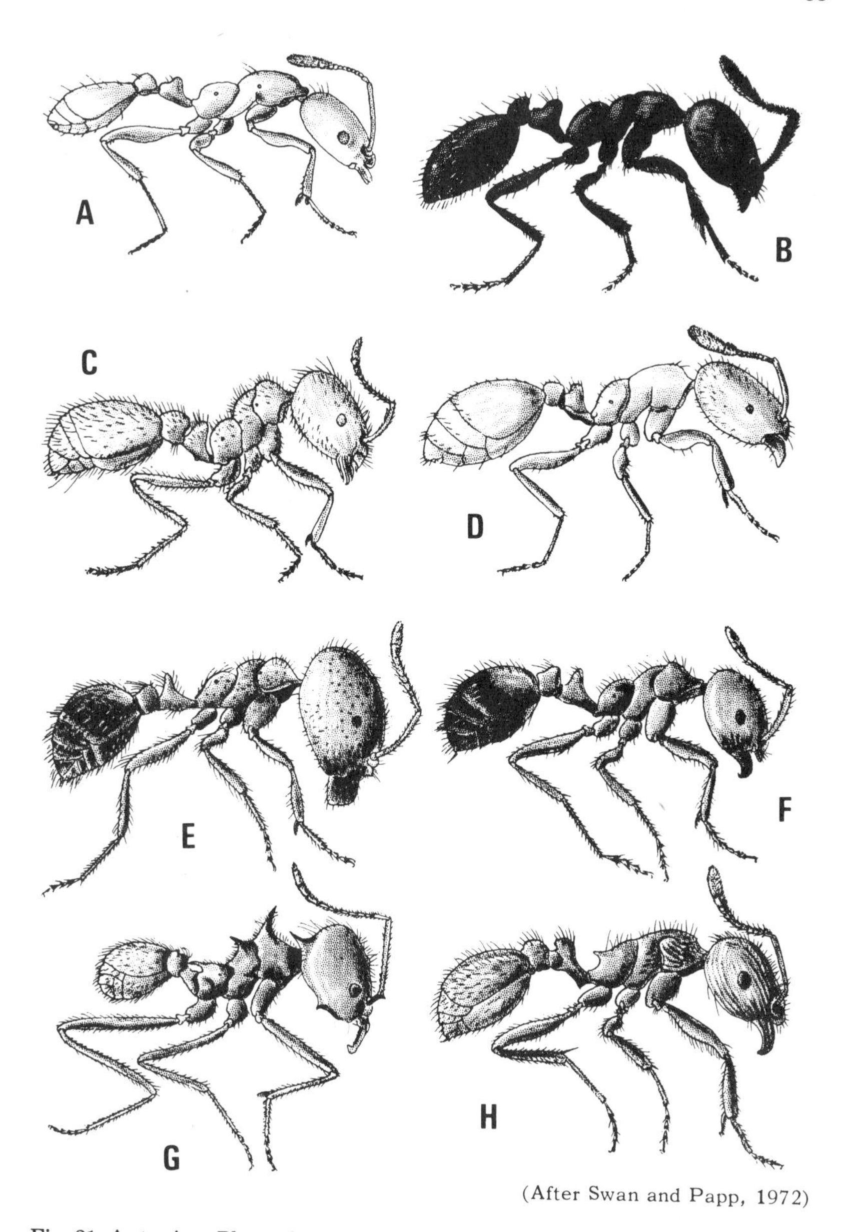

(After Swan and Papp, 1972)

Fig. 21. Ants: A — Pharoah ant *(Monomorium pharaonis);* B — little black ant *(Monomorium minimum);* C — southern fire ant *(Solenopsis xyloni);* D — thief ant *(Solenopsis molesta);* E — fire ant or tropical fire ant *(Solenopsis geminata)* (see also Fig. 19); F — imported fire ant *(Solenopsis richteri);* G — Texas leaf-cutting ant *(Atta texana);* H — pavement ant *(Tetramorium caespitum).* It will be noted that these ants have a two-segmented petiole or waist.

jointed antennal club and petiole, extremely small, from 1 to 1.5 mm long. It nests in moist grassy places and under rocks, is common indoors where it eats meats, grease, cheese, but no sweets. Females of the eastern form are reported to bite and sting people in bed at night. Great mating fights have been

Fig. 22. Top: Aerial view of western harvester ant mounds on infested range, Shoshoni, Wyoming. Bottom: Active mound of the western harvester ant (photos courtesy R.J. Lavigne, Univ. of Wyoming).

observed in Los Angeles and Davis, California in June and July. It is often confused with **Pharaoh's Ant** *(Monomorium pharaonis),* which has a three-jointed antennal club.

The ponerine ants are carnivorous, live in small colonies and nest in the soil or rotten wood. *Ponera ergatandria* occurs in California, Arizona, New Mexico, Colorado, Texas, Florida, south to Central America and the West Indies; it may be recognized by the dull brownish yellow color, narrow petiole (one-segmented), rather small size, being from 3 to 3.75 mm long, the extremely small eyes (with two or three facets), and short antennal scape. In California Massis states that he found this species under rocks in a very moist situation near Devil's Gate Dam, Pasadena, and *P. trigona opacior* around the foundations of homes in Bakersfield. Winged forms of *P. ergatandria* have been reported as so numerous in some fields in Florida as to annoy farm workers "like mosquitoes"; at night they passed through ordinary window screens with ease.

In the small group of pseudomyrmecine ants — intermediate between the primitive ponerine ants and the myrmecine ants — there are several species known to sting severely. One of the most common and widespread of these is *Pseudomyrmex pallidus,* which ranges from North Carolina west to California, south to South America and the West Indies; specimens were collected in San Diego and Imperial Counties in California by beating willows. It is smooth and shiny, pale yellow, head and thorax sometimes slightly reddish, with sparse yellow hairs mostly on gaster, slender two-segmented petiole, from 3.5 to 4 mm long; they are larger and more common in Mexico and southward. The venom gland is very similar to that of the **Imported Fire Ant;** the venom, however, is strikingly different in being proteinaceous, rather than nonproteinaceous as in the case of the fire ants whose venoms have been studied. The worker of *P. pallidus* bites into the skin with its mandibles prior to stinging as do the fire ant workers; it arches the thorax, bends the abdomen underneath, which no doubt helps to drive the lancets into the skin. The sting is followed by a burning pain, reddening of the skin, and later a wheal.

Fire Ants

The name "fire ant" is applied to four species, all limited to the southern part of the United States. The **Imported Fire Ant** *(Solenopsis richteri* and *S. invicta),* which range from North Carolina and Florida west to Arkansas, Louisiana, and Texas (not including Tennessee); the **Fire Ant** *(S. geminata),* which ranges from Texas to South Carolina, and Florida; and the **Southern Fire Ant** *(S. xyloni),* which ranges from California to the southern part of South Carolina and northwest corner of Florida. The **Fire Ant** has often been confused with the **Southern Fire Ant** and has been reported from California, apparently in error; both are pictured here for comparison (Fig. 21).

The **Southern Fire Ant** is a native polymorphic species, with body smooth and shiny, hairy (especially the gaster), head and thorax yellowish red, abdomen black, ten-segmented antennae with two-segmented club, two-segmented petiole, from 1.6 to 5.8 mm long. The **Fire Ant** may be distinguished by the monstrous head, and the narrow petiole with sharply pointed node. The **Southern Fire Ant** is an extremely annoying pest. It disfigures lawns by piling up mounds of dirt, and is very sensitive to ground vibration caused by persons walking over the nest, which is apt to bring ants scurrying from the nest entrance to sting them about the ankles and legs. They are fierce predators and useful in helping to control insect pests, but their bad habits may outweigh their good qualities from man's viewpoint. Their eating habits are extremely variable: beside insects (dead as well as alive) they eat seeds, sometimes damage seed beds and potato tubers, girdle nursery stock, tend homopterous insects for their honeydew, and invade homes where they are attracted to foods of high protein and fat content such as meats, grease, butter and nuts. Their stings are very painful and the reaction severe in sensitive persons. They swarm from May to September but winged forms are seen before and after.

Fig. 23. Imported fire ant mound on East Texas pasture land (photo Agr. Ext. Serv., Texas A&M Univ.).

Carpenter Ants

Carpenter ants *(Camponotus)* are large ants, mostly black, with one-segmented petiole as in other formicine ants, 12-segmented antennae without club. They usually nest under rocks, in stumps, tree trunks, building timbers and utility poles; galleries usually start where the wood has decayed and extend below and above the ground. The workers are polymorphic, the colonies usually large. Carpenter ants will bite but not sting humans. Some species smell strongly of formic acid and aggravate their bites by applying the substance (which is not a venom) to the wound.

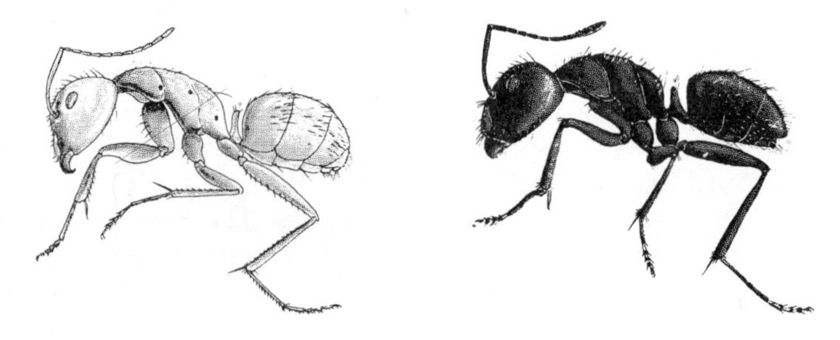

Fig. 24. Carpenter ants: A — brown carpenter ant *(Camponotus castaneus)*; B — black carpenter ant *(Camponotus pennsylvanicus)*.

The **Giant Carpenter Ant** *(Camponotus laevigatus)* occurs from Montana and New Mexico to British Columbia and Mexico, is common at altitudes from 4,000 to 1,000 feet; localities in California include Yosemite, Sierra Nevada Mts., San Jacinto Mts., Baldy Peak in the San Gabriel Mts., Santa Cruz Mts., Alta Peak (Sequoia National Park). It is all black, from 6 to 10 mm long (queens are from 13 to 15 mm long). Large colonies are found in dry stumps or logs; the **Ant Cricket** *(Myrmecophila oregonensis)* is a common guest. *C. sansabeanus vicinus* ranges from Alberta to New Mexico, west to British Columbia and California; in its many varietal forms it is widely distributed in both the lowlands and mountains of California. It is reddish brown and black, from 6 to 9 mm long, nests under stones in dry sunny places. *C. sansabaenus maccooki* occurs from Oklahoma and Texas west to Oregon and California, is common in California around Sacramento and the San Francisco Bay region. It is yellowish to yellowish brown, from 6 to 12 mm long, nests under rocks in moist situations. *C. herculeanus modoc* ranges from South Dakota to New Mexico, west to British Columbia and California, where it is common at elevations from 3,000 feet up. It nests in stumps and rotted tree butts, sometimes tunnels in log cabins, usually on the sunny side, is black with reddish legs, from 6 to 10 mm long. It is a vicious biter when disturbed.

Ant Stings

Fire ants bite and sting simultaneously. They appear to bite as a means of getting a firm hold on the skin while they insert the stinger, or they may use it as a pivot while they sting more than once; there is also the possibility that the wound made by the mandibles serve as a means of inserting the stinger more readily. The alarm in an ant nest which prompts them to leave quickly and in great numbers is no doubt a pheromonal phenomenon as in the case of bees. The **Southern Fire Ant** worker is normally slow-moving when compared with some ants, but responds quickly and moves fast when the alarm sounds overhead. The stinging apparatus of ants is similar to that of bees and wasps, consisting of a sheath called the gorgeret, which encloses two stylets; a duct from the venom gland and the accessory gland enters the base of the gorgeret. As in bees, the two glands provide acid and alkaline components of the venom. [The vemon of *Pseydomyrmex pallidus* is alkaline, but the contents of the accessory gland (Dufour's gland) is neutral (8).] "In stinging, the pointed gorgeret is thrust into the skin and then the stylets are alternately pushed deeper into the wound beyond the tip of the gorgeret . . ." (100).

The venom of fire ants *(Solenopsis)* that has been studied is nonproteinaceous; that of other stinging ants studied is proteinaceous and similar to that of bees and wasps. The **Imported Fire Ant** "alone exhibits necrotic activity as evidenced by characteristic pustule formation at the sting site"; it is described as resembling the necrosis caused by the bite of a brown recluse spider (32). The proteinaceous venom of pseudomyrmecine ants causes an elevation "surrounded by a narrow red halo" to develop about 24 hours after the sting occurs; a depressed discolored area forms in another 24 hours and may last for two

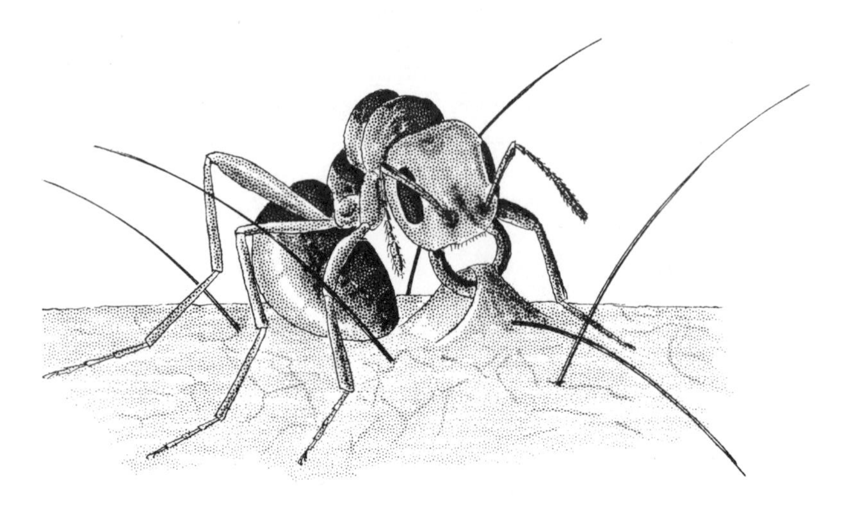

Fig. 25. Fire ant in act of biting. It usually stings simultaneously, using grasp on skin as pivot.

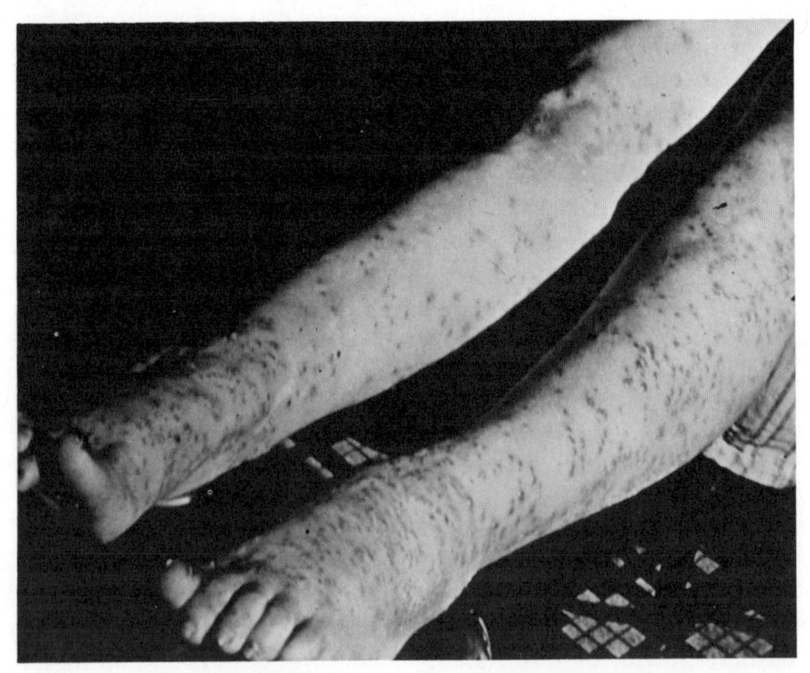

Fig. 26. Stings of the imported fire ant (USDA photo).

weeks. The stings of fire ants and harvester ants are sometimes serious. Smith cites the case of a small infant "reported to have been stung to death" by the **Southern Fire Ant** (84). It is stated by Young and Coppock that "several deaths have been reported from stings of Texas harvest ants in Oklahoma and numerous deaths have occurred throughout the Southwest. These deaths have, in most instances, been infants that crawled onto the nest, or people hypersensitive to stings" (106).

Regarding the stinging habits of harvester ants, Wheeler wrote: "the smaller and more timid species are no more formidable than other Myrmicine ants of the same size dwelling in small colonies. But this is not the case with *P. occidentalis* (**Western Harvester Ant**), *P. barbatus* (**Red** or **Texas Harvester Ant**) and the allied varieties and subspecies. The sting of these ants is remarkably severe, and the fiery, numbing pain which it produces may last for hours. On several occasions when my hands and legs had been stung by several of these insects while I was excavating their nests, I grew faint and almost unable to stand. The pain appears to extend along the limbs for some distance and to settle in the lymphatics of the groin and axillae. If it is true, as has been reported, that the ancient Mexicans tortured or even killed their enemies by binding them to ant-nests, *P. barbatus* was certainly the species employed in the atrocious practice" (100).

First Aid and Prevention

In the case of multiple stings or where sensitive persons experience a severe reaction, first aid as described for bee stings will help prior to medical treatment (see page 26). Ants close to the house or otherwise troublesome should be controlled; the first step is to be sure that food is not left around to attract them. To destroy a colony, the entrance to the nest must be found first. Insecticidal dusts or granules are the easiest and safest to apply, and are simply poured down the hole, and washed down with water. In orchards the insecticide is spread around the trunk at the base of the tree, where the ants pass as they are about to climb the tree for the honeydew feast and on their return to the nest. A bait consisting of soybean oil as an attractant, corncrib grits as a carrier, and an insecticide has been used against harvester and other ants. It is scattered near the entrance to the mound or nest; the worker ants carry the bait into the nest, and in due time the colony is destroyed. The trouble with baits of course is that they attract other animals than the target and sometimes kill birds and other "friendly" creatures. Low concentrations of certain agents, it is claimed, will work against the ants with little or no harm to other forms.

IV. FLIES

Flies (order Diptera, meaning two wings) are characterized by a single pair of wings, instead of two pairs as in other winged insects, and a greatly enlarged thorax (actually an enlargement of the mesothorax or middle segment, the other two being compressed). The wings are attached to the mesothorax, as are the forewings of insects having two pairs. Flies have a pair of knob-like projections called halteres on the metathorax, the last or basal thoracic segment, in place of the hindwings of other insects; they are thought to act as balancers. The order of true flies also include the insects commonly known as gnats, midges and mosquitoes. The mouthparts of flies are very complex. Among the blood-sucking flies, they fall into two types: those with flattened blade-like stylets, as found in the horse flies and deer flies, and those with piercing type stylets, as found in the stable fly and in mosquitoes. These piercing parts are poorly developed in the males. Females of the blood-sucking species need a blood meal for egg production; they are the biters. Both males and females visit flowers for nectar (which is carbohydrate); females need this for the energy necessary for sustained flight. In this group are most of the insect vectors or carriers of microorganisms causing disease in humans and other animals.

Biting Midges

The biting midges (family Ceratopogonidae) are very small flies, most of them from 1 to 2 mm long, and are commonly known as no-see-ums, punkies, moose flies and sand flies. They are sometimes mistaken for the nonbiting midges (family Chironomidae) or the black flies. They are short, chunky flies compared to the slender true midges. A distinctive feature is the peculiar wing venation, and the fact that the wings are relatively

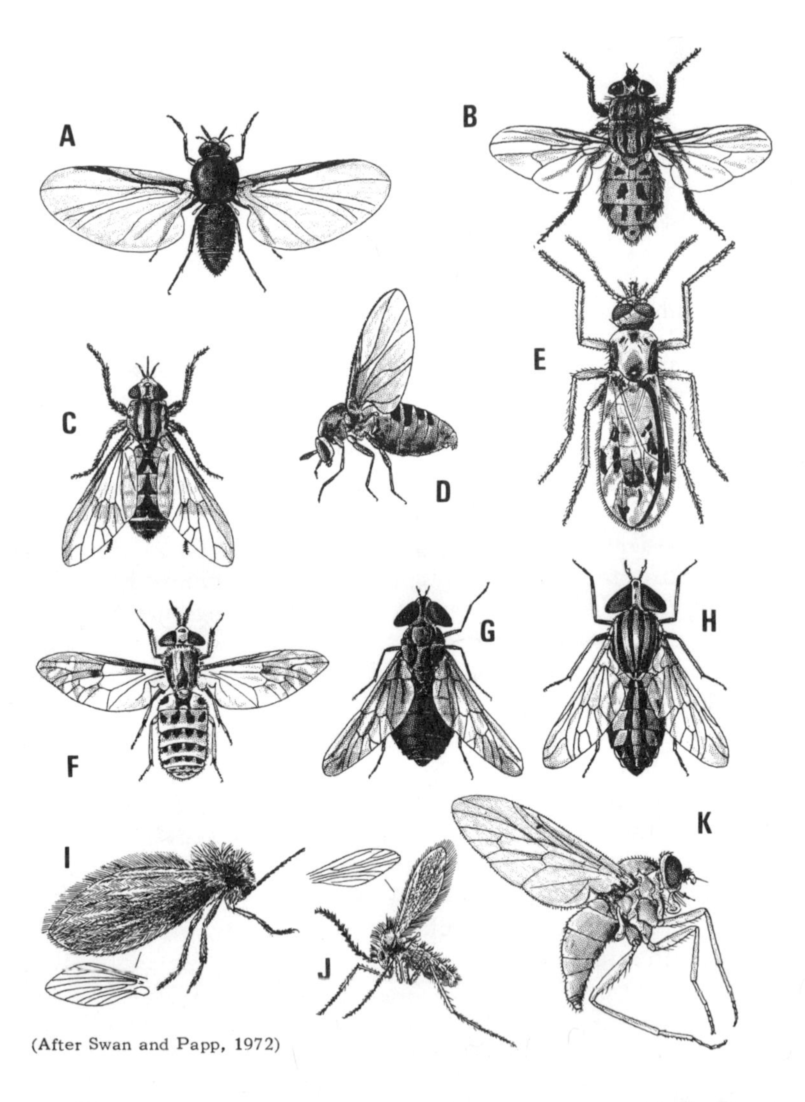

Fig. 27. Flies: A — black fly *(Simulium venustum)*; B — false stable fly *(Muscina stabulans)*; C — deer fly *(Chrysops callidus)*; D — black fly *(Simulium vittatum)*; E — biting midge *(Culicoides furens)*; F — deer fly *(Chrysops discalis)*; G — black horse fly *(Tabanus atratus)*; H — striped horse fly *(Tabanus lineola)*; I —moth fly *(Psychoda alternata)*; J — sand fly *(Phlebotomus)*; K — snipe fly *(Symphoromyia atripes)*.

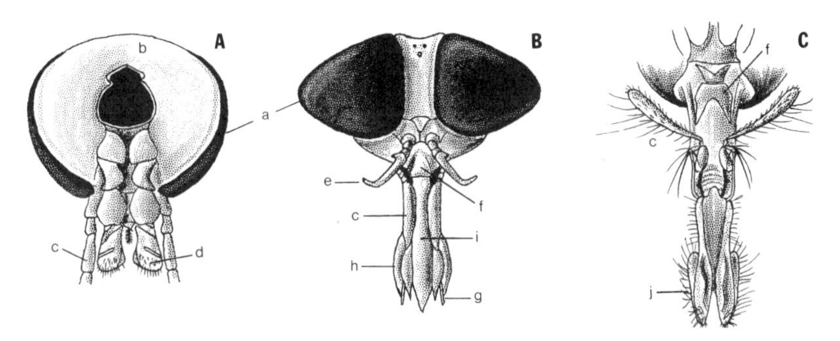

Fig. 28. Mouthparts of some biting flies: A — black fly *(Simulium);* B — horse fly *(Tabanus);* C — bottle fly *(Calliphora).* In details: a - eye; b - occiput; c -maxillary palpus; d - paraglossa; e - antennae; f - clypeus; g - maxilla; h - labium; i - labrum; j - labellar lobe.

large and nearly always folded flat over the back when at rest. The mouthparts are well developed for piercing and sucking, the proboscis being long and having six blades enclosed. Biting midges breed in aquatic and semiaquatic environments: at the edge of streams, lakes and ponds, in salt marshes or water-holding plants, in mud and other damp places. Most species are predaceous on other small insects, some suck blood from the wing veins of large insects, among them dragonflies. The blood sucking pests of man and other vertebrates are found in the genera *Culicoides, Leptoconops,* [The scientific names of these gnats may change since the genus is undergoing revision at this writing.] and *Lasiohelea* (which is not western). They bite savegely, the favorite areas of attack being around the hat band and collar. Locally they may be so numerous at times as to make a place virtually uninhabitable. Rabbits, and birds especially, are among the wild hosts; cattle, horses, sheep and poultry among domesticated animals are attacked. Biting midges have been associated with the transmission of filarial worms and viruses causing disease in animals and man.

The two most common species of *Leptoconops* are the **Bodega Black Gnat** *(Leptoconops kerteszi)* and the **Valley Black Gnat** *(L. torrens).* Both are vicious biters and attack in bright sunlight. The **Bodega Black Gnat** appears black, but is actually brown when viewed under the microscope, the head and thorax being dark brown and the abdomen a much lighter brown; the wings are transparent but when folded on the back appear milky white in the reflected light. The female is from 2.5

to 3 mm long, the male from 3.5 to 4 mm long, and lighter in color. It is found on the margin of coastal salt marshes and saline lakes throughout California, and is said to be even "a worse pest around some of the alkaline and salt lakes in Utah" (19); it also occurs in North Africa (Egypt and Tunisia). The gnat was named after Bodega Bay in California where it was first studied. The **Valley Black Gnat** closely resembles the **Bodega Black Gnat;** the females may be distinguished by the antennae, which are 14-segmented in the former and 13-segmented in the latter. The males of the **Valley Black Gnat** are slightly smaller than those of the other species; they occur in California in the great Central Valley, Santa Clara Valley and Riverside County and are also found in Colorado and Arizona.

The **Bodega Black Gnat** breeds in the sand around the edge of brackish pools, lakes and marshes. The **Valley Black Gnat** breeds in clay-adobe type soils (such as we have in the great valleys), which crack when dry and swell when wet. In the spring or early summer, after a blood meal, the female crawls down a crack in the soil to lay her eggs. The larvae pupate here the following year, after which the new generation of flies emerge from the fissures in the soil. Irrigated orchards often serve as a breeding ground. Besides man the black gnats feed on chickens, turkeys, dogs, cats, rats and probably any other warm-blooded animal available. Until they have had their blood meal the gnats are vicious and persistent and cannot be denied or brushed aside.

Only three California species of *Culicoides* "have a bad reputation as biters," according to Wirth and Stone (96). There is little comfort in this since what they lack in number of species they seem to make up for in numbers of individuals. The bad biters are: *obsoletus,* found in the northern mountains, *tristriatulus* along the northern coast and extending to Alaska, and *reevesi* in the hilly parts of the lower San Joaquin Valley. *C. obsoletus* is a small shiny brown species, with wings nearly bare, excepting a few hairs at the apex, one dark and two lighter spots on front margin of wing, extensive shadings on rest of wing, from 1 to 1.5 mm long; besides northern California it occurs in the three far western provinces of Canada, north to Alaska. *C. tristriatulus* is a large dark brown species, with faint dark stripes on thorax, about 2.3 mm long; it is a serious pest along parts of the Alaskan coast. *C. reevesi* is a very small brown species, about 1 mm long, resembling *obsoletus;* it has been reported from Arizona and New Mexico as well as California.

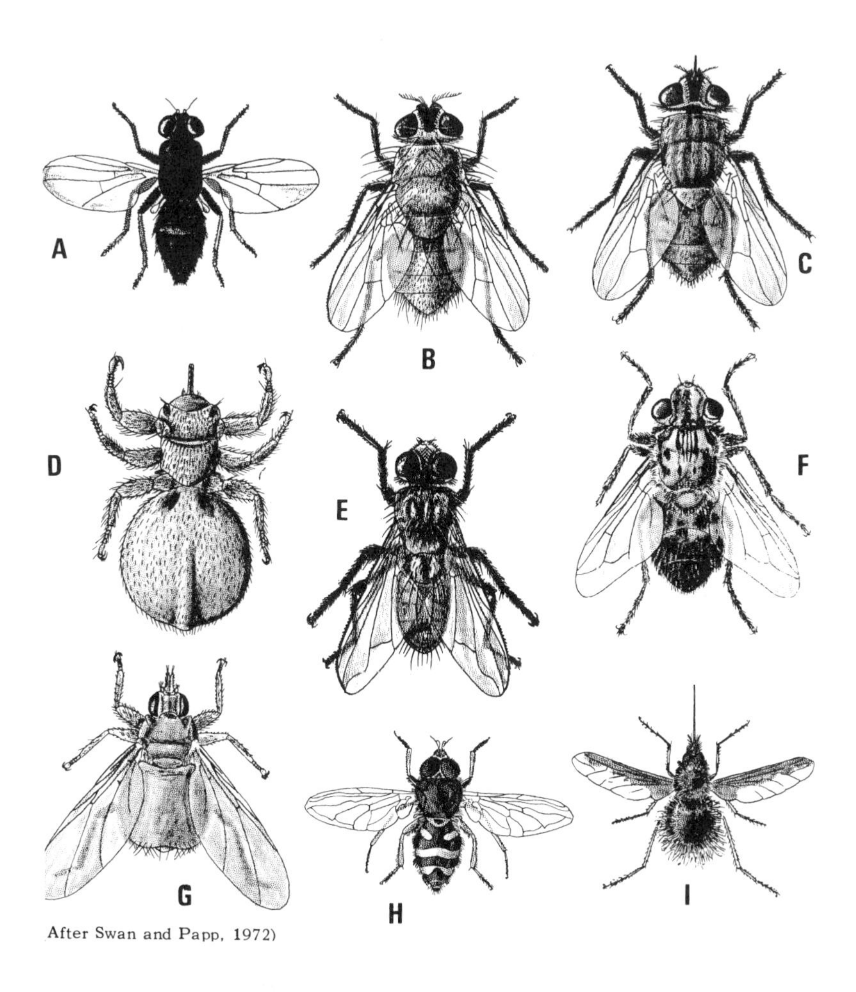

Fig. 29. Flies: A — eye gnat *(Hippelates pusio);* B — false stable fly *(Muscina assimilis);* C — stable fly *(Stomoxys calcitrans);* D — sheep ked *(Melophagus ovinus);* E — house fly *(Musca domestica);* F — human bot fly *(Dermatobia hominis);* G — louse fly *(Lynchia americana);* H — flower fly *(Syrphus torvus);* I — bee fly *(Bombylius major),* a non-biting fly resembling a bee (the larva is a parasite of wild bees).

Black Flies

Black flies, also known as buffalo gnats and turkey gnats (family Simuliidae), are stout-bodied little flies with their heads pointing downward, giving them a humpbacked appearance; the wings are short and broad, with anterior veins heavy. They are not always black; many are grayish brown or yellowish. The male's eyes are very large and close together; the upper facets are much larger than the lower ones and they are divided by a horizontal line. The female's eyes are distinctly separated by a frontal stripe. The ocelli are absent in both sexes. Black flies are extremely abundant along streams in forested areas of the temperate zone. The immature forms are aquatic. The larvae and pupae attach themselves to rocks at the bottom of swift-flowing streams in moss-like masses; they are common in irrigation canals and flumes, sometimes in such great numbers as to impede the flow of water. The females have short grasping type mouthparts and bite without warning. They are most active in the morning hours, swarming about the head and biting with aggravating persistence. The bites cause swelling and itching. Black flies torment domestic and wild animals as well as humans, and sometimes kill livestock. They are intolerable in some areas in the spring and early summer unless one wears a net over his head or uses an effective repellent. The simuliids have not been implicated in the spread of human diseases in North America, north of Mexico.

The **Striped Black Fly** *(Simulium vittatum)* is black, with gray pollinose coating, mesonotum with five dark brown stripes, abdomen with three indefinite rows of black spots sometimes fused to form a central black stripe, the male darker than the female, from 2 to 3 mm long. It is one of the most common and widely distributed black flies, ranging from the Arctic to Mexico, and from the Pacific to the Atlantic, in California from Modoc County to Monterey and San Diego counties. It attacks domestic animals and occasionally man. *S. bivittatum* is pale in color, with two broad silvery pollinose stripes on the mesonotum, dark lateral spots on abdomen faint or missing, a small species about 1.7 mm long, occurs from Alberta and Saskatchewan south to California, and eastward to South Dakota, Nebraska and Texas, in California from Monterey and Fresno counties to

San Diego County. Cole calls them "wicked little biters" (19). *S. trivittatum* is yellow, mesonotum with three dark brown to blackish stripes, abdomen with lateral spots nearly as dark as the median spots, about 2 mm long; it occurs from Mexico and California to New Mexico and Texas, in California from Butte County to Santa Barbara and San Bernardino counties. *S. griseum* is almost completely pale yellow or yellowish gray, thorax with practically no pattern or only a faint median stripe, male with mesonotum metallic green and sometimes a rather distinct median stripe, about 1.7 mm long; it ranges from Saskatchewan and Alberta to California, Colorado and Texas, in California in San Bernardino, Riverside and Imperial counties.

The **Turkey Gnat** *(Simulium meridionale)* is grayish to brownish black, with thick gray dusting, mesothorax with three narrow black or brown stripes, from 2 to 3 mm long; it occurs from Alaska to Indiana, south to California, Florida and Mexico, in California in Butte County, is a tormentor of man and domestic animals. "This is the turkey gnat of the Mississippi Valley," where it is a serious pest of turkeys, chickens and pheasants and is known to transmit a blood parasite to turkeys. *S. notatum* is a pale yellowish brown species close to *griseum* but distinguishable from it by the legs being all yellow excepting the fore tarsi and apical segments of the middle and hind tarsi, antennae all yellow; it is a very small species, from 1 to 1.5 mm long, found in California, Arizona, Nevada, New Mexico and Texas, where it is troublesome to man and domestic animals, sometimes occurring in great swarms on the latter. *S. venustum* is shiny black, thorax with short yellowish pubescence, abdomen with sparse brown pubescence, from 2 to 2.5 mm long, occurs throughout North America, from Alaska to Greenland and southward, in California from Plumas County to El Dorado County; it is an aggravating pest of man and animals. *S. arcticum* is black, two basal segments of antennae yellowish, mesonotum with two large white pollinose spots, a large species, 3 to 4 mm long; it occurs from Alaska to California, east to Manitoba, Utah, Colorado and Arizona, in California from Siskiyou County to Riverside County. About 40 species of black flies are listed by Cole as occurring in western North America (19); more than half of these are found in California according to Wirth and Stone. Most of the species annoying to man are in the genus *Simulium*. "There is no evidence that any of the California species (of *Prosimulium)* are annoying as pests of man, as is *Prosimulium hirtipes* in the North and East" (96).

Horse Flies and Deer Flies

Horse flies and deer flies (family Tabanidae) have been called "gadflies" because of the extreme annoyance of their sudden, vicious bites. They are medium to large in size, stout-bodied, with powerful wings and large heads (composed mostly of eyes). They are generally black or brown in color, often striped or spotted and are found around lakes, swamps, streams, salt marshes, beaches and irrigated land. Eggs are laid in masses on vegetation near or over the water, and drop from the plants; the predaceous larvae are mostly aquatic or semiaquatic. The blood-sucking adults usually attack only the larger animals, including man. The females have sharp blade-like stylets which are inserted deep in the flesh, causing blood to flow. The blood is taken up by the sponge-like labellum at the end of sheath enclosing the stylets; the loss of blood as a result of these wounds is quite considerable among cattle and has a debilitating effect.

Horse flies are believed to be mechanical vectors of bacteria causing anthrax (that is, they carry the organisms but the infection is introduced into the host through a cut or scratch). Also known as malignant pustule or carbuncle, and wool sorter's disease, anthrax is an infection of domestic animals

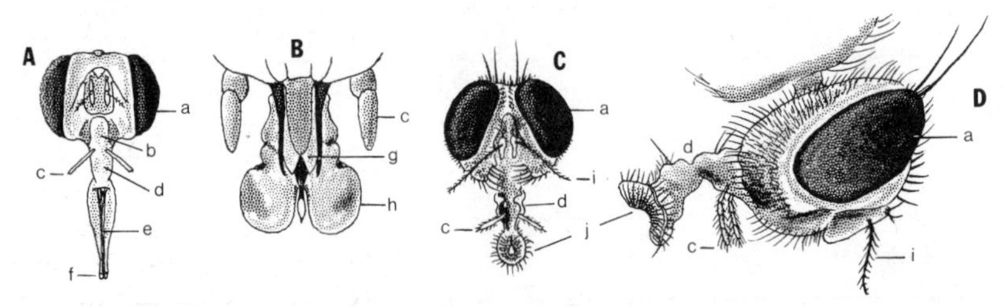

Fig. 30. Mouthparts of some biting and nonbiting flies: A — stable fly *(Stomoxys);* B — snipe fly *(Symphoromyia),* cutting-sponging type, the mandibles acting as blades and the labium as spone; C — house fly *(Musca domestica),* nonbiting, sponging type, front view; D — same, side view. In details: a - eye; b - clypeus; c - maxillary palpus; d - rostrum; e - labrum; f - labellum; g - mandible; h - labium; i - arista; j - labella.

and other mammals including man; reservoirs of infection are rodents and rabbits. Some deer flies are mechanical vectors of bacteria causing tularemia (also known as deer fly fever), an infectious disease of man and other mammals; reservoirs of infection are rodents and rabbits, more especially the latter.

The majority of horse flies are in the genus *Tabanus,* and deer flies in the genus *Chrysops,* at least in their present taxonomic state. In *Tabanus* the ocelli are absent, the third antennal segment has five divisions and the hind tibiae are without apical spurs. In *Chrysops* the ocelli are present, the antennae are more slender than in *Tabanus,* the third segment with five divisions, and the hind tibiae have two apical spurs. One of the best known members of the first group is the **Black Horse Fly** *(Tabanus atratus),* a widespread species recorded in the West from Idaho, Wyoming, Utah and Colorado, but not from the Pacific coast; it is a huge fly, about 25 mm long, jet black, thorax with whitish or yellowish pubescence, abdomen often with bluish cast, wings black to smoky brown, unspotted. The **Western Horse Fly** *(T. punctifer),* which occurs throughout western North America and eastward to Texas and Oklahoma, is no less formidable, being about the same size as *atratus,* black, with cream-colored pile on the thorax, and a small brown spot on the wing.

The **California Horse Fly** *(Tabanus californicus)* ranges from British Columbia to Shasta and Santa Cruz counties in California, is brownish, with sides of abdomen orange, quite large, from 17 to 19 mm long. *T. frontalis* [*T. septentrionalis* Loew is now a synonym of *T. frontalis* Walker.] is widespread from Alaska to Utah and east to New York, grayish black, thorax with grayish stipes, abdomen with median row of yellowish gray triangles and sublateral row of brownish orange spots, from 13 to 16 mm long. *T. opacus* occurs from Alberta to northern California, Utah and Colorado, is grayish brown, abdomen with median row of yellowish gray triangles and sublateral row of yellowish oblique stripes, a small species, from 12 to 15 mm long. *T. sonomensis* ranges from Alaska and Alberta to California, Colorado and Wisconsin; it is black, with sides of abdomen orange near base, from 12 to 15 mm long. *T. rhombicus* occurs from Alaska to Oregon, Colorado and Arizona; it is a "small, stout, blackish" species, with three rows of gray triangles on the abdomen (faint in some specimens). Horse flies are not the persistent biters that deer flies are, but their occurence in large numbers, especially at bathing beaches, often make them extremely annoying pests. Among the western species, *T. punctifer, frontalis, sonomensis,* and *rhombicus* are noted for this. Fifty some species of *Tabanus* in all are listed as western by Cole (19).

The **Deer Flies** are medium in size, seldom over 12 mm long, have brilliant green or golden eyes with zigzag stripes, and distinctive wing markings which are useful in identification of species. They are usually found in damp woods or swampy areas and are well known to trout fishermen, campers and hikers as persistent biters. *Chrysops callidus* is perhaps the best known and most widespread member of this group, ranging from British Columbia to Maine, and southward to Colorado in the West; it is black, with yellowish markings, callus (a lump) below the eyes yellow, abdomen with yellow middorsal triangles and pale yellow spots on sides near base, from 7 to 9 mm long. *C. discalis* ranges from British Columbia and the Canadian prairie provinces through the western states, the Dakotas and Nebraska; it is light grayish, thorax with three greenish gray stripes (stripes wider and black in the male), abdomen with four rows of black spots, wings with discal cell clear and small spot at fork of third vein, from 8 to 10.5 mm long, a vector of bacteria causing tularemia in rabbits and man.

Chrysops excitans is transcontinental in distribution, ranges across Canada and the northern United States from New England, New York and New Jersey west to Alaska, Oregon, Washington and California (Plumas, Shasta and Sierra counties); it is extremely variable, head yellow pollinose, thorax black, with dense yellowish pubescence and faint grayish stripes, abdomen black, with yellow middorsal triangle often on second segment, sometimes on others, from 7 to 12 mm long. *C. noctifer* occurs from California to Washington, Oregon and British Columbia, also in Nevada and New Mexico; it is mostly black, abdomen with yellow spot on each side near base, from 8 to 9.5 mm long. *C. proclivis* is a northern species, found in Alaska and Alberta, ranging south to Colorado and northern California; it is predominantly black, thorax with yellowish pubescence, abdomen mostly black, first segment yellowish, with black quadrate spot, from 7.5 to 9 mm long. *C. coquilletti* occurs in California and Utah; it is prodominantly yellow, first two segments of abdomen with black mid-spot, third to fifth segments each with four black spots, remaining segments black with yellow posterior margins, a robust species from 8 to 9 mm long. A total of 29 species of *Chrysops* are recorded as western (19), 19 of these in California (96).

The Stable Fly and Its Close Relatives

The **Stable Fly** *(Stomoxys calcitrans)* resembles the closely related **House Fly** *(Musca domestica)* — they belong to the family Muscidae — but it is a little larger, about 8 mm long, and may be separated from most other flies by the long slender, horny proboscis. The dark round spot on the abdomen, and the slight upward curve of the third wing vein also serve to distinguish it from the **House Fly.** It is cosmopolitan in distribution, having "followed man and his domestic animals all over the world." The **Stable Fly** is a vicious biter and attacks domestic animals and man; the males suck blood with the females and clothing does not ordinarily afford humans the protection they might expect. The proboscis is admirably adapted for its function of piercing and "is carried like a bayonet ready for charge; it is forced into the victim by a strong thrust of the head and body" (32). The fly is not known to be a vector of organisms causing disease in humans in this country. Among horses it is a mechanical vector of trypanosomes causing surra, and is implicated in the mechanical transmission of a virus causing infectious anemia.

The **False Stable Fly** or **Nonbiting Stable Fly** *(Muscina stabulans),* which is widely distributed in North America, is easily confused with both of the above flies. The abdomen is dark gray, the tibiae yellowish; it is about 8 mm long, has four dark longitudinal stripes on the thorax, like the **House Fly,** but the third wing vein is bent only slightly upward, whereas it meets the second vein near the apex in the **House Fly.** It feeds on milk and other liquid foods, lays its eggs in decaying animal and vegetable matter. *Muscina assimilis* closely resembles the **Nonbiting Stable Fly** and has similar habits; European in origin, it is widely distributed across the United States and southern Canada and occurs along the Pacific coast.

The **House Fly** is, of course, not a biter (it has sponging type mouthparts) but because of the similarity it is often blamed, along with its nonbiting close relatives, for the bites of the

Stable Fly. The **House Fly** is abundant everywhere; because of its association with filth and close association with man, it is considered the greatest threat to human health of any single species of insect. It is about 6 mm long, the thorax gray, with four darker longitudinal stripes, the abdomen gray or yellowish, with darker median line and irregular yellowish spot at the anterior lateral margins; the arista (the terminal portion of the antennae) is plumose. The straw-colored spots left on surfaces are regurgitated food, the dark spots are fecal matter; by this means and by direct contact with food, they spread typhoid fever, dysentery, diarrhea, cholera, pinworm, hookworm and tapeworms.

The **Little House Fly** or **Lesser House Fly** *(Fannia canicularis)* [Some authors place the genus *Fannia* in the family Anthomyiidae.] is also common, cosmopolitan in distribution, and closely associated with man; it ranges from Alaska and Greenland to California and Florida, completely replaces the **House Fly** in many subarctic regions. "It may be found in the West from sea level to mile-high canyons in the mountains" (19). It is similar to the **House Fly** in appearance, but smaller, from 3 to 5 mm long, and more slender, dull gray, thorax with three darker longitudinal stripes, abdomen with yellowish areas on sides; the third wing vein continues straight to the apical margin, and the arista is not plumose. The **Latrine Fly** *(F. scalaris)* occurs in all states of the United States and the Canadian provinces; it is 5 to 7 mm long, the thorax dark grayish, with four brownish stripes, the abdomen gray pollinose, with median black stripe. It breeds in dung, garbage, human excrement; in nests of bees, wasps and birds, rodent burrows and dead snails and is attracted to lights. It is known to cause gastric and intestinal disturbances (myiasis) due to the presence of the fly maggots. *F. benjamini,* first named from specimens found in California, is widely distributed north to Idaho and Wyoming, south to Mexico, and east to Arkansas, brownish black, face and lateral margins of mesonotum grayish-dusted, palpi, halteres and legs yellow, abdomen with central black stripe, round black spot on side of each segment and whitish dusting, wings clear with yellowish veins, from 3 to 3.5 mm long; it is attracted to human perspiration and mucous discharges from the eyes, and is a nuisance in some places.

Snipe Flies

The snipe flies (family Rhagionidae) are small to medium-sized, long-legged flies, usually with stout bodies and pointed abdomens. They are sluggish flies and breed mostly on land. Some species in the genus *Symphoromyia* (and in the related Neotropical genus *Suragina*) have the blood-sucking habit and are extremely annoying to large mammals, including man. The females light on the human skin quietly, usually singly and most often on the hands, and inflict a painful bite. These curious flies are not elusive like most other flies and can be poked around with the finger without causing them to fly away, and they can be easily picked up. The males are not attracted to animals and are usually found on foliage. *Symphoromyia atripes,* described by Aldrich as "the bad biter *par excellence,"* is widespread from Alaska to California and Colorado; it is mostly black, thorax grayish, with four wide darker stripes, abdomen all black, with dense pile, wings smoky grayish brown, from 5 to 8 mm long (the male a little larger than the female).

Symphoromyia plagens, said to be "a persistent biter," occurs from British Columbia to California, at sea level in the north and along mountain streams in southern California. The head and thorax are black, overlaid with yellowish pollen, abdomen chestnut-brown, antennae, palpi, femora and tibiae yellow, wings mostly clear; the female is 5.5 to 7 mm long, the male 8 mm long. *S. limata* is found in the mountainous regions of southern California where it is abundant and irritating to trout fishermen; it is a large black species, with antennae, palpi and tibiae yellow, abdomen black on first four segments, red on the remaining segments, wings brownish, about 8 mm long. *S. pachyceras,* described as "a vicious biter," has been reported from Oregon, Washington and Arizona and is common in north and central California; it is almost wholly black, the female from 6 to 9 mm long, the male about 6 mm long. About 15 species of *Symphoromyia* occur in the West and in California, but only a few are reputed to bite.

Eye Gnats and Louse Flies

Eye gnats belong to a family of flies (Chloropidae) with exceedingly varied habits, running the gamut from a pest of wheat (the fruit fly) to gallmakers, predators and vectors of disease organisms. Eye gnats (or *Hippelates* flies) are very small, chunky flies, dark gray to black in color, which breed in soils having high organic content. They are common in warm, dry regions, where they dart about the eyes, nose, mouth and exposed wounds of animals with great persistence, in search of moisture and mucous; secretions of the latter provide the female with the protein necessary for egg production. They do not bite the way blood-sucking flies do; the end of the mouthparts has spines with which the gnat grasps the mucous membranes, thereby adding to the possibilities of spreading disease organisms mechanically. They are believed to cause the spread of pinkeye (conjunctivitis) infections among humans, and bovine mastitis among cattle. In the tropics they are suspected of being involved in the spread of yaws.

Hippelates pusio is widespread from Washington to North Dakota and Pennsylvania, south to California, Florida and Mexico, and is considered a serious pest in the southern states where it annoys dogs and people; it is dark gray to black, about 1.5 mm long. The closely related *H. collusor* (which may be the same species) is "the abundant and troublesome species in southern California," where it was implicated in an epidemic of pinkeye in a school; it also occurs in Nevada, Arizona and Mexico. *H. hermsi* is black, with yellow forelegs, found in California and Nevada, east to New Mexico and Texas. *H. pallipes* is widespread from North Dakota to Quebec and Maine, south to Utah, California, Texas and Florida. About 12 species of *Hippelates* occur in the West, 8 of them in California.

Louse flies (family Hippoboscidae) are strange-looking flies — flat and leathery, the wingless ones looking more like blood-sucking lice or some forms of ticks than dipterans. They are external parasites of birds and mammals and rarely leave their hosts. Those found on birds usually have wings, most of those found on mammals do not, which would seem to be a convenient adaptation to their modes of living. They are unusual among insects in being born as mature larvae, ready to pupate or change to the adult form, a phenomenon known as pupiparity. Probably the best known of these insects is the **Sheep Ked** *(Melophagus ovinus),* which is widespread from British Columbia to Maine, south to California, Florida and South America. It has short thick legs with toothed and hooked tarsal claws, only rudimentary wings reduced to knobs which are sometimes mistaken for halteres (the latter are absent), and is about 6 mm long. The **Sheep Ked** is normally found only on domestic sheep, but it is not uncommon for them to get on sheephandlers and inflict a painful bite. The **Pigeon Fly** *(Pseudolynchia canariensis)* is an Old World species that has emigrated to the New World, and has become established from California to Wisconsin and Massachusetts, south to Texas, Florida and South America. The thorax is shiny brown, abdomen brown, legs yellowish brown, wings pale yellowish; it is about 6.5 mm long on the average. "In the New World *P. canariensis* is strictly a parasite of domestic pigeons . . . flies often into man and will bite him spontaneously" (5). It is damaging to squab pigeons, especially as a vector of "pigeon malaria."

Mosquitoes

Mosquitoes (family Culicidae) are the most important insects affecting health of man, with the exception of the **House Fly,** since they include the vectors of organisms causing many serious communicable diseases: yellow fever, malaria, encephalitis, filariasis and dengue. [Mosquitoes were the chief vectors of the virus causing a serious outbreak of Venezuelan equine encephalomyelitis in Mexico and Texas in 1971. More than 11,000 horses died and many people became ill from the disease, which is relatively mild in humans and seldom fatal.] They are small frail insects with long legs and scaly wings, the latter being distinctive for the complete marginal vein. The females, which are the biters and have the bloodsucking trait, possess a remarkable instrument admirably adapted for the purpose (and our discomfort and that of many animals) as is apparent from our drawing of the mouthparts. The males have poorly developed mouthparts — they live on the nectar of flowers — and may be distinguished from the females by their bushy antennae. The piercing-sucking mouthparts of the female consist of six stylets: four (mandibles and maxillae) for cutting and two (hypopharynx and labrum-epipharynx) which when pressed together form a tube for conducting the blood. Two pumps operated by muscles of the pharynx provide the suction. Secretions of the salivary glands which cause the itching are injected into the host through a smaller duct in the hypopharynx.

The most important mosquitoes are in the genera *Aedes, Anopheles,* and *Culex,* but there are some large common species in the genus *Culiseta,* and a few fierce biters and troublesome species in the genus *Psorophora.* The eggs of *Aedes* mosquitoes are spindle-shaped and scattered over the soil; they hatch when flooding takes place. The eggs of *Culex* mosquitoes are capsule-like and laid in clusters or "rafts" on the water; those of *Anopheles* are boat-shaped, pointed at both ends, and are laid

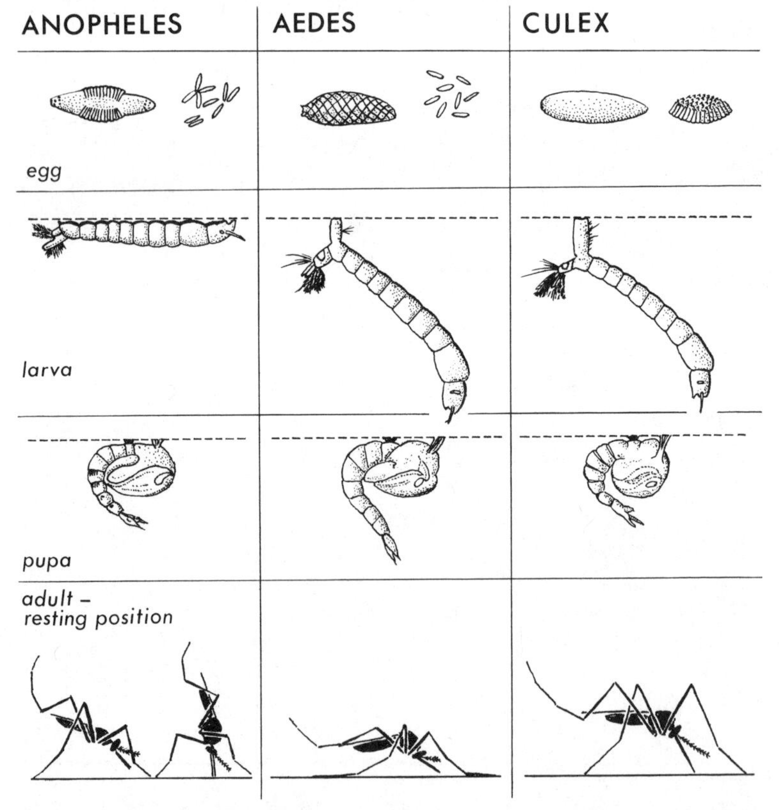

Fig. 31. How to distinguish the most important groups (genera) of mosquitoes: *eggs* — shape and manner of dispersal, as shown in first series across the page; *larvae* — position when at surface of water for air; *adult females* — resting position when not engorged or hibernating.

singly on the water. *Aedes* and *Psorophora* adult females may be distinguished nearly always from those of other genera by the tapered end of the abdomen, and the two short cerci protruding from the end. Most adults of *Anopheles* may be recognized by their resting position; the abdomen and proboscis are held in nearly a straight line, pointed at an angle toward the resting surface. *Culex* adults rest with the body parallel to the surface. The palpi of *Anopheles* are as long as the proboscis in both sexes, and enlarged or club-shaped in the males; in *Culex* the female palpi are short, those of the male longer than the proboscis.

The **Dark Rice-Field Mosquito** or **Florida Glades Mosquito** *(Psorophora confinnis)* is said to have caused the death of 80 cattle and 67 swine in Florida in two days. It is common in the southeastern states and the rice-growing areas of Arkansas and Louisiana, but is found much farther north, from Massachusetts to Colorado and occurs in the alfalfa-growing areas of the Imperial Valley in California. A medium-sized, sooty black mosquito with white-ringed proboscis and tarsi, pale diffuse apical scaling or patches on abodmen, and wing length 4.5 mm., it breeds in temporary pools, attacks at night or in shady or grassy places during the day. The **Alaska Mosquito** *(Culiseta alaskaensis)* is one of the large "snow mosquitoes," which hibernate as adults and emerge in the spring while much of the winter's snow is still on the ground. It occurs in the mountainous areas of Utah and northern California, northward to Alaska and Siberia. The mesonotum has an even mixture of brown and white scales, which sometimes form a pair of median stripes, the abdomen is black with white band at base of segments, legs black with wide white basal bands, wings with dark scales forming spots. The **Rainbarrel Mosquito** *(Culiseta incidens),* which probably doesn't find many rainbarrels anymore, occurs in the West from Alaska to California where it is one of the commonest mosquitoes. It is a dark brown species, mesonotum with mixture of dark brown and yellowish scales, abdomen black with white basal segmental bands, legs dark brown with narrow white rings, wings with dark scales forming patches. Also known as the **Cool Weather Mosquito,** it breeds throughout the year in forested areas where the temperature is not too low, attacks horses and cattle more than man and is a vector of equine encephalitis.

The **Saltmarsh Mosquito** *(Aedes sollicitans)* is widely distributed from Florida to Arizona, north to the Dakotas, has golden brown thorax, abdomen with white or yellow basal segmental bands and white median line, proboscis and tarsi with broad white bands, wing speckled with white and brown, about 4 mm long. A fierce biter that attacks man and domestic animals in broad daylight, it breeds in salt marshes and areas intermittently flooded with salt water. It is a strong flier, migrates in large swarms long distances from any marsh. The **Black Saltmarsh Mosquito** *(Aedes taeniorhynchus)* occurs in the Atlantic and Gulf coast states and along the Pacific coast from southern California to Peru. It is black, abdomen with broad white segmental bands, proboscis and tarsi with sharply contrasting white bands, wing entirely dark scaled, about 3.5 mm long. Its habits are similar to those of *sollicitans;* the females are hard, persistent biters and "appear at times in

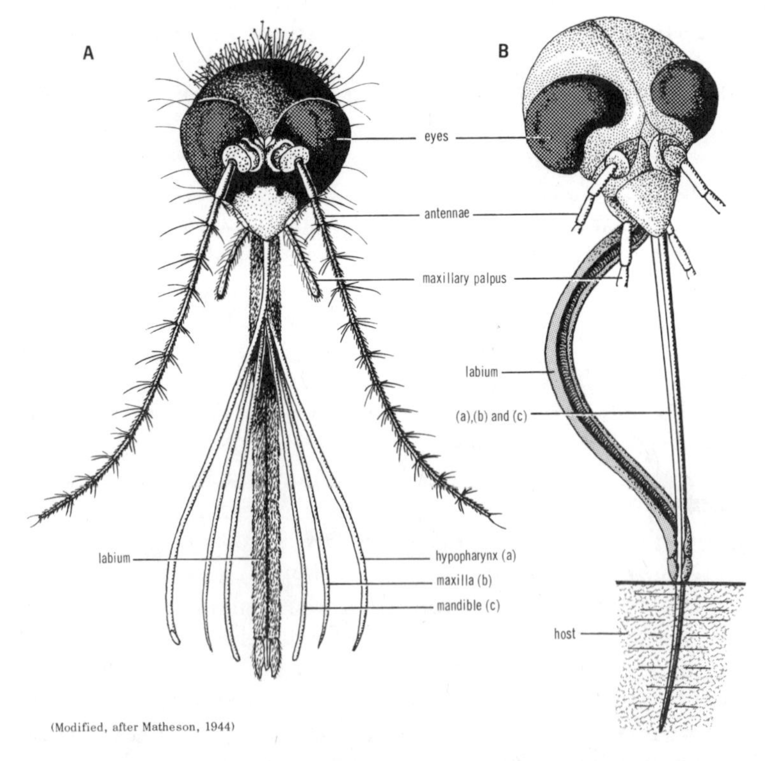

Fig. 32. Mouthparts of female mosquito: A — front view of head with mouthparts spread apart; B — stylets inserted in host.

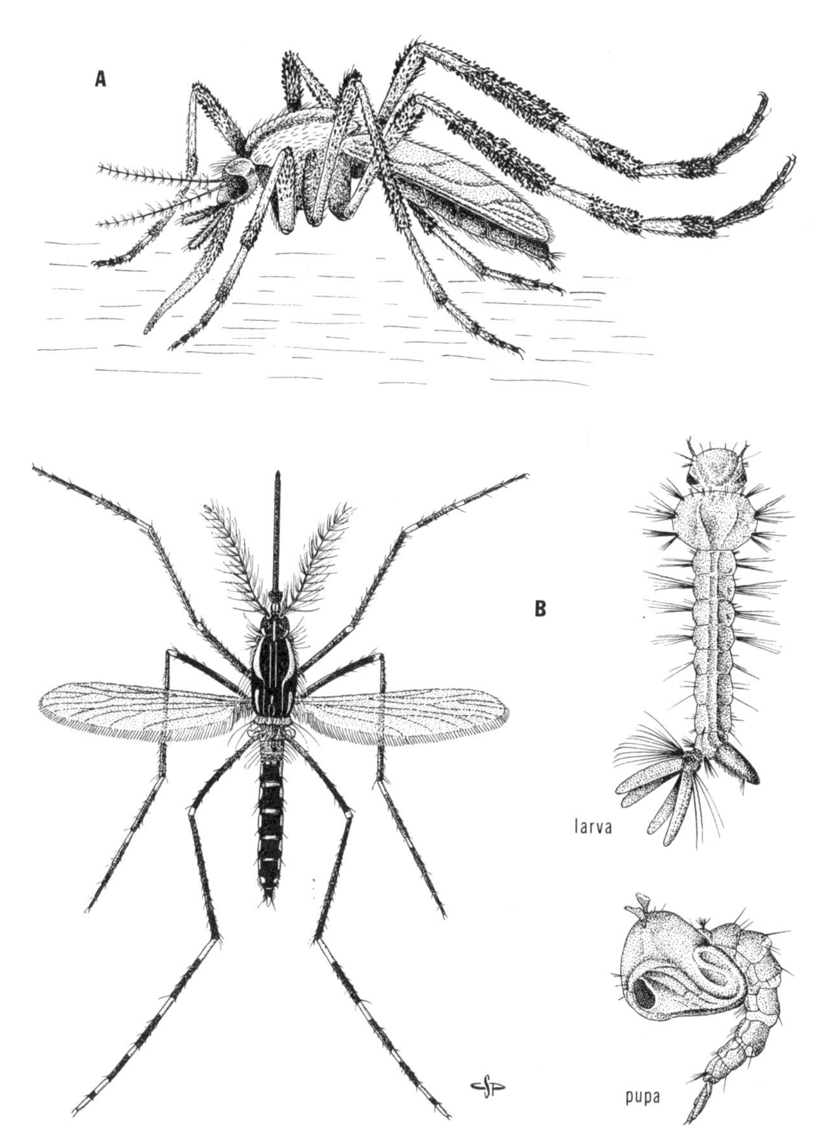

Fig. 33. Mosquitoes: A — gallinipper *(Psorophora ciliata)*; B — yellow fever mosquito *(Aedes aegypti)* adult, larva and pupa.

unbearable numbers." The **California Saltmarsh Mosquito** *(Aedes squamiger)* is a large brownish gray species with white-ringed tarsi, breeds in salt marshes along the Pacific coast; a common species overlapping *dorsalis* in the Bay region of

California. The **Brown Saltmarsh Mosquito** *(Aedes dorsalis),* [In "Common Names of Insects" (Ent. Soc. Amer.) an eastern species, *Aedes cantator,* is named **Brown Saltmarsh Mosquito.** This name has long been associated with *A. dorsalis* on the Pacific coast.] widely distributed in northern and western United States and Canada, is yellowish gray, head, proboscis, palpi and legs speckled with white, mesonotum with diffused brown median stripe, abdomen pale-scaled with white, line of dark paired quadrate spots, wing pale-scaled and speckled with black, about 4 mm long. It breeds in brackish and fresh water along the northern Pacific coast but extends far inland; it is the dominant species in many places in the West, and is a vector of virus causing encephalitis.

The **Floodwater Mosquito** *(Aedes sticticus)* occurs in the northern and western states and Canada under flood conditions, and is found in the Sacramento Valley of California, along with *vexans.* The mesonotum has a broad central stripe of brownish scales, usually divided by a fine median line of yellowish scales, abdomen with white bands, widening laterally, at base of segments, tarsi not ringed, wings dark-scaled, about 4 mm long. The females are fierce biters, often migrate several miles; they are at times a serious menace to man and animals. The **Vexatious Mosquito** *(Aedes vexans)* is widespread in North America and other continents, has abdomen with broad white bands notched on posterior borders, narrow white basal rings on tarsal segments, thorax and proboscis unmarked, wings about 4 mm long. Primarily a floodwater breeder, it is the principal pest mosquito in some northern areas, but no more vexatious than many of its cousins.

The **Northwest Coast Mosquito** *(Aedes aboriginis)* occurs from Washington to Alaska and Saskatchewan. The mesonotum is yellowish to golden brown with paired dark brown stripes, abdomen black with white basal segmental bands expanding laterally, legs black, wings almost always dark-scaled. A rather large mosquito, it breeds in ditches and "foul pools in the deep forests where the females bite day or night." the **Common Snow Mosquito** *(Aedes communis)* is very similar to *aboriginis;* the mesonotum has dull yellow or gray scales with pale median line separating a pair of dark brown stripes, abdomen dark-scaled with patches of pale scales at base of costa. It breeds in pools formed by melting snows in mountain meadows and forests. A fairly large mosquito, it is common and a fierce biter, as summer visitors to the higher altitudes will testify.

The **Western Malaria Mosquito** *(Anopheles freeborni)* is common in the arid regions west of the Continental Divide; along the coast and in the Sierra Nevada Mountains it is displaced by *A. occidentalis.* A medium-sized blackish species, *A. freeborni* may be distinguished by the "four characteristic dark scale spots on the wing"; it breeds in sunlit pools, and is the common malarial vector of the West. In *A.occidentalis* the wings are dark scaled, and have a silver fringe spot.

The **Northern House Mosquito** *(Culex pipiens)* ranges from the Atlantic to the Pacific in southern Canada and the United States, excepting the extreme South. The proboscis, tarsi and wings are dark, the thorax is pale brown- or grayish-scaled, thc abdomen has broad white basal segmental bands continuous with lateral spots, their posterior margins fairly straight, wing about 4 mm long. It is regarded as the most abundant night-flying mosquito of the northern states, breeds in any standing water. This is the mosquito implicated in outbreaks of St. Louis encephalitis; it is also a known vector of filarial worms. The **Southern House Mosquito** *(Culex quinquefasciatus)* is found in tropical and subtropical regions, occurs in the southern states, westward to California. It is very similar to *pipiens,* may be distinguished by the smoothly rounded posterior borders of the segmental bands, narrowly separated from the lateral spots; the wing is 4 mm long, with scales long and slender. Like *pipiens* it breeds in containers and other places with standing water. The **Common Mosquito** *(Culex tarsalis)* ranges from Canada to Mexico in the West, eastward to Indiana and along the Gulf Coast; it is abundant in the rice fields of the Central Valley of California. A dark brown or black species, with abdominal bands of white scales continuous with lateral spots, it may be recognized by the wide white ring on the middle of the proboscis, the wide basal and apical rings on the hind tarsal segments, and similar but narrow rings on the front and middle tarsi; the wing is about 4 mm long. A dusk and night time biter, it enters houses readily and is considered "the most certain vector of western equine encephalitis." About 75 species of mosquitoes are recorded as western (19), and about 35 of these in California (96).

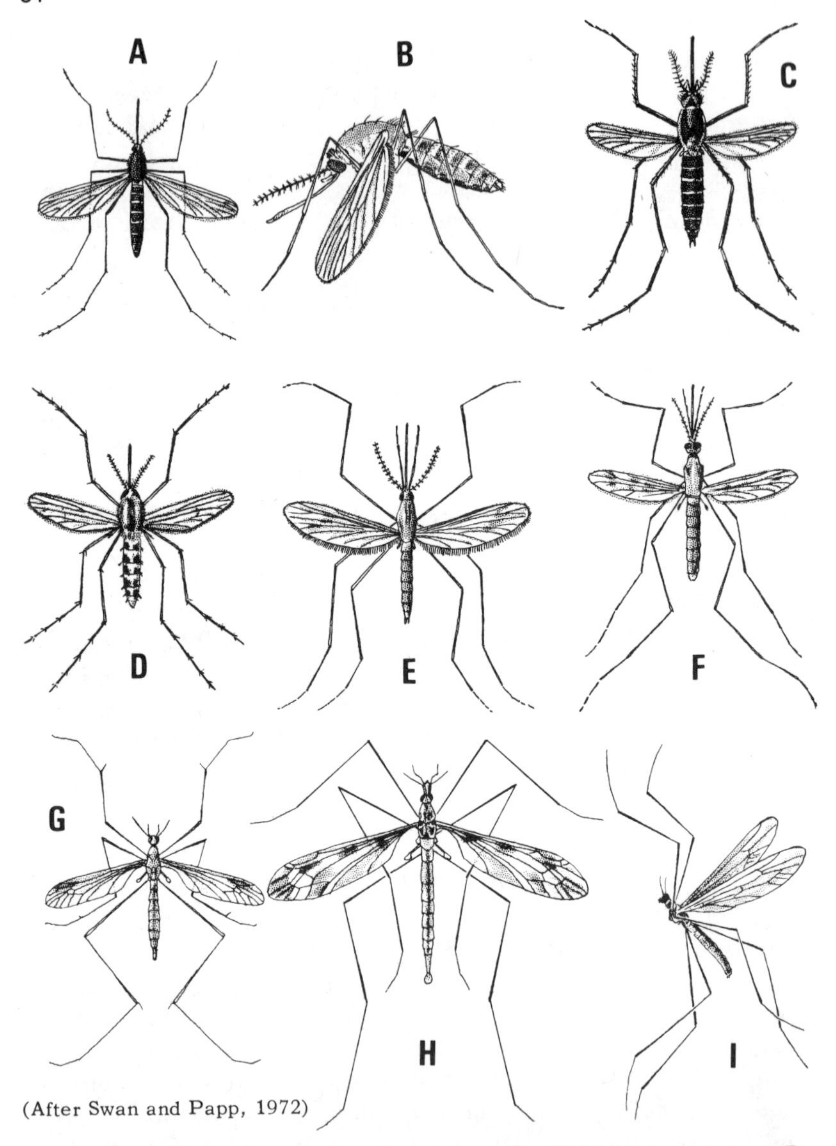

Fig. 34. Mosquitoes: A — northern house mosquito *(Culex pipiens pipiens);* B — side view of same, more enlarged; C — black saltmarsh mosquito *(Aedes taeniorhynchus);* D — saltmarsh mosquito *(Aedes solicitans);* E — common malaria mosquito *(Anopheles quadrimaculatus);* F — western malaria mosquito *(Anopheles freeborni).*

Crane flies: G — a crane fly *(Tipula cunctans);* H — a crane fly *(Tipula trivittata);* I — European crane fly *(Tipula paludosa).* Crane flies shown for comparison; they are nonbiting and generally much larger than mosquitoes but are often mistaken for them.

Nature of Fly Bites

The reaction to bites (swelling and itching) is due to salivary gland secretions injected by the fly when it pierces the skin and before it withdraws blood. The fluid is an anticoagulant which reacts on the hosts' blood and "prepares" it, as it were, for ingestion and later digestion by the insect. It is well known that some people are more attractive to flies than other persons. This has been proven conclusively with mosquitoes in the laboratory, and is borne out in clinical studies. Most of us have experienced the situation where one person is under attack while another in the same place and at the same time is not bothered by the mosquitoes, or other flies. The reaction to mosquito bites in sensitive persons may be immediate or delayed, or both, and is followed by the appearance of wheals and reddening of the skin. Whether due to toxins or histamine components of the salivary secretions injected by the mosquito, and mechanical damage to the tissues, or other causes, is not clearly understood. Natural desensitization after being subjected to mosquito bites for a time has been noted. Some persons develop seasonal immunity after a short initial period of exposure in the spring and become less sensitive to the bites. [Dr. Emile Van Handel of the Florida State Health Board stated (1963) that "nothing does the human heart more good than a sharp bite by a black fly or mosquito." He contended that since the substance injected during the bite is a powerful anticoagulant which prevents clotting and coronary obstruction, it might account for the lower incidence of heart disease in tropical countries, where mosquitoes are more abundant than in regions having more efficient insect control.]

Black flies are probably the severest biters despite their size. It is believed that when they bite "a local anesthetic is injected, because the first thing noticed may be a drop of blood." Not until an hour or so later does the reaction of swelling, pain and itching appear, followed by small blisters or hardened lumps. Frazier states that "biopsy (microscopic examination) has shown histologic (tissue) changes in the skin as long as seven months after the black fly bite. Death, which occurs occasionally, may be due to toxemia or anaphylaxis," that is, blood poisoning or excessive sensitivity to the introduction of protein (32).

The biting midges do not appear to inject secretions having anesthetic properties as the black flies are believed to do, since the bite is felt almost immediately after the proboscis is inserted in the skin, more especially so in the case of *Culicoides*. The reaction to *Culicoides* bites is usually less intense than to those of *Leptoconops*. In sensitive persons they cause itching followed by welts which may persist for several days. Smith and Lowe note that the bite of *Leptoconops* is usually not felt during the first 30 to 60 seconds of feeding (83). Shortly after this a red spot appears on the skin at the site of attack, later followed by nodular swelling and intense itching in some persons a blister may form. In sensitive persons the itching may last for days and even weeks, and can lead to secondary infections as a result of scratching. The reaction to the bites of deer flies and the **Stable Fly** are similar to those described for mosquitoes, but because of their greater size the bites are generally more painful for the moment at least. As with other insect bites, the reaction of individuals varies from mild to extreme.

Repellents are available which, when applied to the skin and clothing, will provide good protection against biting flies. As in the case of fleas, mites and ticks, preparations containing adequate amounts of diethyl toulamide are the most effective. (For a discussion of repellents, see page 140). According to biochemist Dr. Emory Thurston, two or three 10-milligram tablets of thiamine hydrochloride (vitamin B^1) taken at mealtime before and during exposure to mosquitoes will repel them ("they dislike the mild odor your skin will exude"), if you are one of those whose body chemistry produces this effect; unfortunately it doesn't work that way for everyone.

V. FLEAS AND SUCKING LICE

Fleas and sucking lice are external parasites of man and other warm-blooded animals; the females require a blood meal to reproduce. Among the primates (man, apes and monkeys) only man is a host of fleas. Sucking lice (order Anoplura) are not closely related to fleas (order Siponaptera), though both have the parasitic trait. The sucking lice and the more closely related chewing or bird lice (order Mallophaga) are more ancient in origin and less highly developed than fleas. They are easily distinguished by the shape of the body, which is flattened top to bottom (dorsoventrally) in lice and compressed the opposite way (laterally) in fleas. The chewing lice are external parasites of birds (all birds are infested with them) and some mammals, but not man; they are easily distinguished from sucking lice by the larger head and definite separation of the thoracic segments, and by the chewing mouthparts.

Fleas

Fleas are very small insects, usually less than 2 mm long, without wings and having well developed mouthparts adapted for sucking, from whence comes the name (Siphonaptera) of the order. They have long, stout legs and a smooth cuticle with backward-directed bristles, enabling them to move swiftly through the host's hairy or feathery coat. Some show a strong preference for certain kinds of hosts but many attack a great variety of animals, including man. Excepting the sticktight flea, most of them transfer readily from one host to another. The eggs are dropped haphazardly and usually fall to the ground or floor, where they hatch. The larvae, which are usually whitish with long sparse hairs, develop in the dirt; they wriggle violently when disturbed.

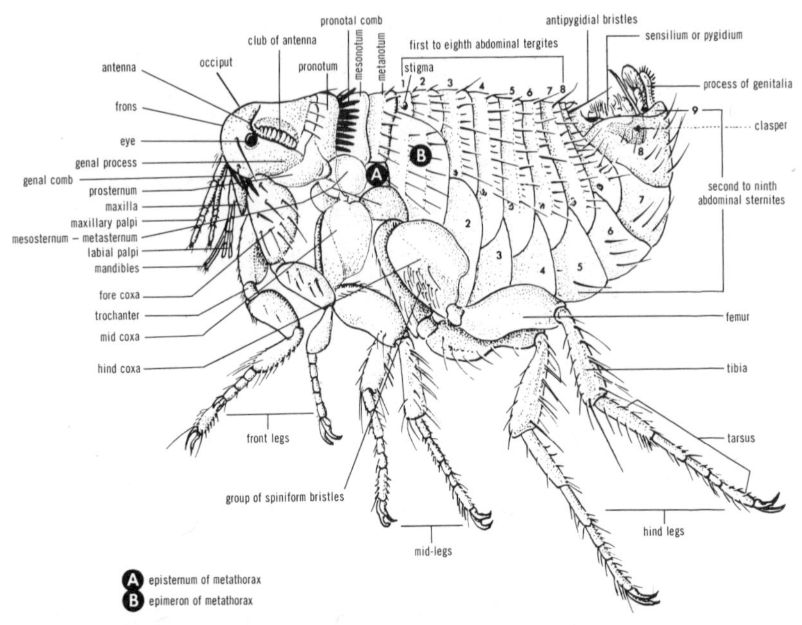

Fig. 35. Detail of dog flea. Sideview, showing head and genal comb, thoracic segments (top — pronotum with comb, mesonotum and metanotum; underside — mesosternum and metasternum, the first segment of prosternum not being visible here), abdominal segments (top — tergites; underside — sternites), leg segments, bristles and other features.

The **Cat Flea** *(Ctenocephalides felis)* and the closely related **Dog Flea** *(Ctenocephalides canis)* are cosmopolitan in distribution and frequently become established in homes where they are a nuisance. Their hosts are cats, dogs, rabbits, rats, squirrels and man. The **Dog Flea** is generally more common in cool climates, while the **Cat Flea** is more abundant in warm climates; the **Dog Flea** is "rather uncommon in California." [Benjamin Keh, Sr. Public Health Biologist, Berkeley, in a personal communication.] The **Cat Flea** may be distinguished by the long flat head, the seven or eight curved black spines (called the genal ctenidium or comb) along each side of the cheek, and the eight spines (the pronotal ctenidium of comb) on each side of the first thoracic segment. The **Dog Flea** is very similar, but may be distinguished by the higher and more rounded head and by the fact that the first genal comb is much shorter than the second one.

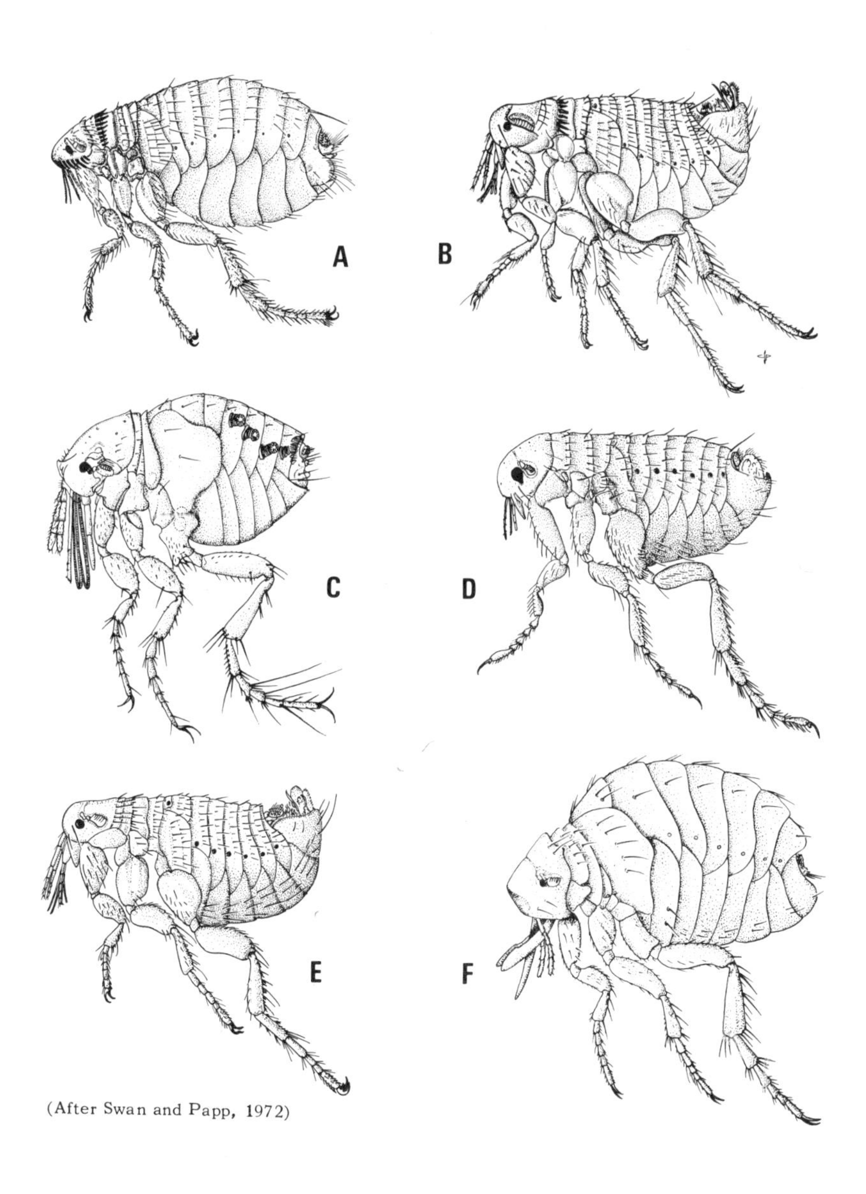

(After Swan and Papp, 1972)

Fig. 36. Fleas: A — cat flea *(Ctenocephalides felis)*; B — dog flea *(Ctenocephalides canis)*; C — chigoe *(Tunga penetrans)*, a "sticktight" flea found in Mexico and southward and occuring on man between the toes and under toenails; D — human flea *(Pulex irritans)*; E — Oriental rat flea *(Xenopsylla cheopis)*; F — sticktight flea *(Echidnophaga gallinacea)*.

The so-called **Human Flea** *(Pulex irritans)* is world-wide in distribution. While it occurs generally across the continent, it seems to be most common in Pacific coastal regions, and is seldom found in large cities. Though it "has adapted itself to residence in folds of man's clothing as a substitute for the fur of the lower animals" (7), it has been recorded from a wide variety of hosts besides man, among them rodents, the pig, dog, cat, opossum and deer. The **Oriental Rat Flea** *(Xenopsylla cheopis),* which resembles the **Human Flea** superficially, should be mentioned because medically, it is "the most important flea in the world" (42), and though primarily tropical, it sometimes becomes established in temperate regions, especially in port cities, and to some extent inland. It attacks man as well as rats and other rodents, and is the principal vector of bacteria *(Pasteurella pestis)* causing plague and of rickettsiae *(Rickettsia typhi)* causing murine and endemic typhus. Both of these fleas lack the genal and pronotal combs as found in the **Cat** and **Dog Fleas;** they may be distinguished from one another by the ocular bristle, which is below the eye in the **Human Flea,** and in front of the eye in the **Oriental Rat Flea.**

The **Sticktight Flea** *(Echidnophaga gallinacea)* is a dark brown species, occurs in the southwestern states and along the Pacific coast as far as Oregon; it is found on poultry, birds, cats, dogs, rabbits, ground squirrels, rats, horses and sometimes humans. The fleas attach themselves in masses on the face, wattles and earlobes of chickens and turkeys and about the ears of cats and dogs; occasionally they get under the skin of poultry handlers. The **Chigoe** *(Tunga penetrans),* another "sticktight flea," is a tropical species which sometimes extends northward from Mexico. The female bores under the skin of its host after mating and lays its egg here, sometimes expanding to about the size of a pea; it attacks poultry, birds, cats, dogs and other animals, as well as man. On poultry the fleas are usually found around the eyes and comb, on dogs and cats they may be found around the ears or feet; they attack man mainly under the toenails and between the toes. The two species are similar in appearance but may be readily separated by the presence of a patch of small spines on the inner side of the hind coxa of the **Sticktight Flea,** and their absence in the case of the **Chigoe** (89).

See also Appendix 1 (page 170) and 2 (page 175).

Nature of Flea Bites

Because of their remarkable jumping ability — the human flea can jump six to eight inches high and a distance of twelve to fifteen inches — fleas (other than the "sticktights") are not caught easily for identification, unless some suction device is used. Bites may occur on any part of the human body but most often on the arms, legs or neck. Small children, because of their close association with pets, are most likely to be the victims. As in the case of mosquitoes, and some other biting insects, some persons are more attractive as hosts than others. Flea bites are usually grouped in an irregular manner, and result in typical papular swellings with eruption of the skin. The reaction, which may be immediate or delayed, is believed to be from oral secretions injected by the flea at the time of taking its blood meal. Humans go through a period of sensitization when subject to flea bites. As the bites occur, sensitization increases and earlier bites may "flare up," causing the typical papular lesions mentioned above. According to Frazier, "With continued exposure to flea bites, man undergoes a natural desensitization to the antigen deposited by the flea" (32). Good protection against fleas may be obtained from repellents, applied to exposed areas of the skin or to the clothing; preparations containing an adequate amount of diethyl toulamide seem to be the most effective against fleas as well as biting flies, mites and ticks. Thiamine hydrochloride (vitamin B[1]) taken orally gives varying degrees of protection from fleas, depending on the person; for some it gives good protection, for others it is of no value (32). For further discussion of repellents see page 140.

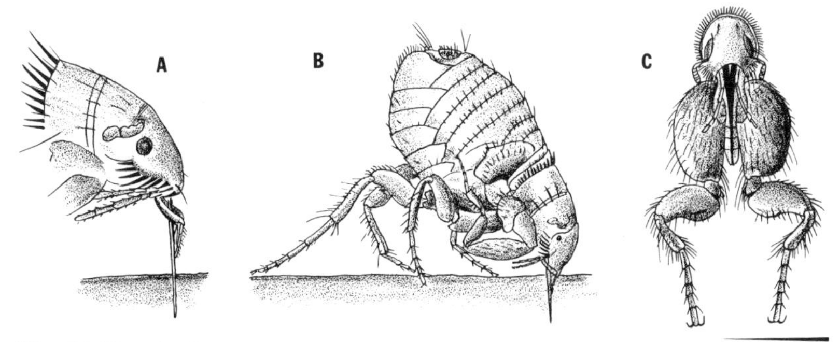

Fig. 37. Flea feeding: A — sucking tube piercing skin; B — sucking tube fully inserted; C — front view of flea.

Sucking Lice

Sucking lice are very small insects, somewhat larger than fleas as a rule, and like them, wingless. Unlike the fleas, they are usually host-specific (that is, confined to one or closely related species) and generally remain on the one host. The lice are transferred from one person to another by direct contact or indirectly by contact with infested clothing or headgear. The eggs, which have a distinct cap, are glued to the hairs of the host, or to clothing fibers in the case of the **Body Louse.** There is no larval stage as in the fleas; the young (or nymphs) look like adults, except that they are smaller. Lice have highly developed mouthparts, modified for piercing and sucking and consisting of three stylets which are retracted within the head when not in use. Most mammals, including seals and elephants (but not bats, nor marsupials) are hosts of one or more species of sucking lice. Animal lice do not infest man. Three kinds of lice infest humans: the **Body Louse** *(Pediculus humanus humanus),* the **Head Louse** *(Pediculus humanus capitis)* and the **Pubic** or **Crab Louse** *(Phthirus pubis).* They are cosmopolitan in distribution.

The **Body Louse** and **Head Louse** are very similar in appearance and may be difficult to distinguish, except that the part of the body where found will often provide a clue. The **Body Louse,** however, is normally much larger than the **Head Louse** — both the male and female are usually more than 2 mm long — and paler in color, with more slender antennae and less pronounced constrictions between the abdominal segments. It

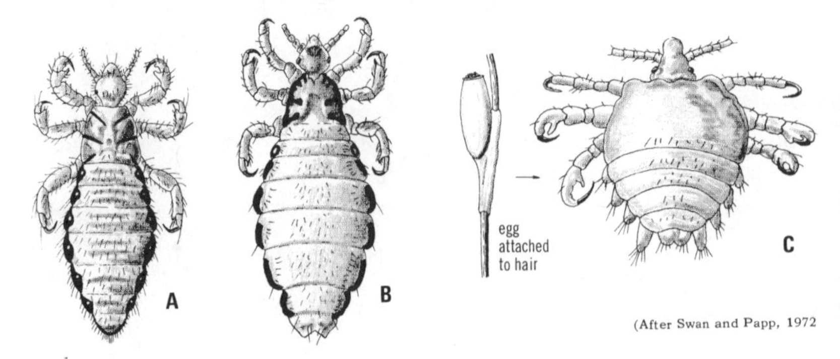

Fig. 38. Human lice: A — body louse *(Pediculus humanus humanus);* B — head louse *(Pediculus humanus capitis);* C — crab louse *(Phthirus pubis).*

is unique in not remaining in contact with the host at all times; when not feeding it rests in the folds of the clothing. If the infested clothing is not worn for several days, all the lice in it will die. The **Body Louse** is the principal vector of rickettsiae *(Rickettsia prowazeki)* causing epidemic typhus — a more serious disease than fleaborne (or endemic) typhus; it is rare in tropical countries where little clothing is worn. Typhus epidemics have usually been associated with crowding and unsanitary conditions resulting from wars. Unlike plague organisms, those of typhus are not transmitted through the bite of the insect but rather through abrasions of the skin or by way of the mucous membranes (eyes, nose or mouth) when the infected louse is crushed. Typhus is a disease of the insect's gut and is transmitted through the feces rather than salivary secretions.

The **Head Louse** is gray in color, normally less than 2 mm long; the male is considerably smaller than the female. It is commonly found in the hair of the head, causing itching and irritation of the scalp, particularly at the back of the head. Infestations may also occur on other hairy parts of the body, such as the beard or mustache; the lice rarely migrate to the pubic region. A heavy crop of hair is more especially susceptible to infestation. The **Head Louse** is not implicated in the transmission of disease organisms. [An alarming increase in **Head Lice** has been noted by public health officials since long hair became a fad among male youngsters.]

The **Pubic Louse** is crablike, whitish, shorter than the other lice found on humans; short hairy processes protrude from the sides. While normally occurring in the pubic region, it is sometimes found on the eyebrows or in the armpits. It can survive only short periods removed from man and is usually spread by personal contact, only occasionally by way of infested toilets, furniture or bedding. The **Pubic Louse,** except possibly in rare cases, does not transmit disease organisms. [The occurrence of **Pubic Lice** was reported (1974) to have reached epidemic proportions in the United States, the incidence being greatest among females 15 to 19 years of age and males over 20.]

Nature of Lice Bites

The reaction from lice bites varies with individuals, as in the case of other kinds of bites and may be immediate or delayed. The **Body Louse** feeds frequently, making a new puncture each time, causing irritation and itching. Tiny red spots appear first on the body and later become expanded into small papules. The salivary secretions of the lice are believed to be the cause of the reactions, but crushing the insect on the skin or the presence of the insect's feces may have a part.

Infestation with lice (referred to medically as pediculosis) commonly occurs under poor hygeinic conditions but anyone may become infested, either by contact with another person or indirectly from bedding, clothing, furniture and washrooms. "Body lice favor older persons and usually spare children," according to Frazier (32). Children, on the other hand, often become infested with head lice and epidemics in schools are not uncommon. The egg stage in the case of the **Head Louse** is the easiest to detect; the eggs, known as nits, may be seen attached to the hair, most often close to the scalp behind the ears.

Persons who practice reasonable sanitation, and bathe and change their clothes frequently need have no fear of body lice. The **Pubic Louse,** which is usually found on adults rather than children, is not so often associated with filth as are other lice found on humans. The nits look much like those of the **Head Louse** and are glued to the hairs. Reaction to the bites may vary from only a slight discomfiture to severe itching, and presence of the lice may not be recognized for a long time. They invariably leave tell-tale bluish spots on the skin which may persist for several months; the spots do not disappear when pressed, as usually happens in the case of red spots on the skin. "Self-treatment for lice may cause serious consequences," according to the U.S. Public Health Service. It is recommended that infested persons consult their doctor or health department.

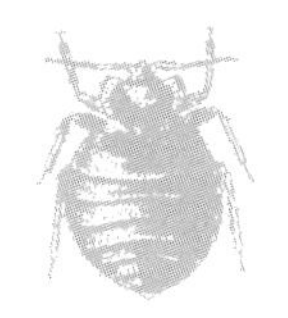

VI. BUGS AND BEETLES

Strictly speaking, the term *bug* applies only to a certain group of insects — the order of true bugs or Hemiptera; the scientific name means "half-wings" and refers to the fact that about half of the forewings is hardened, and the other half membranous and thus functional in flying. We have applied the term here in its restricted sense.

Assassin Bugs

The assassin or reduviid bugs (family Reduviidae) as a whole benefit man by preying on insect pests, but some of them are vicious biters while others (in the genus *Triatoma)* are bloodsucking and can transmit organisms causing disease in man. The **Western Bloodsucking Conenose** *(Triatoma protracta)* is found in the nest of wood rats but often enters houses, trailers and tents where it frequently bites humans asleep in bed. It is found throughout California, ranges east into Utah and south into Mexico. The bug is black in color, 16 to 19 mm long, has a *straight,* tapered proboscis or beak which pivots into a groove in the underside of the head and forepart of the thorax (64). The beak is equipped with four fine hollow stylets, one of which pierces the skin when the proboscis is extended for bloodsucking. Besides man and wood rats, the bug will attack dogs, cats, opossums and armadillos.

The **Western Corsair** *(Rasahus thoracicus)* — a common reduviid in California, Arizona and Mexico, and frequently seen at lights — makes a squeaking noise when handled. The bug is 4 to 6 mm long, amber in color; the wing membranes are black, with a large amber spot near the middle of each forewing. It is mainly nocturnal in habit, and preys on the larvae and

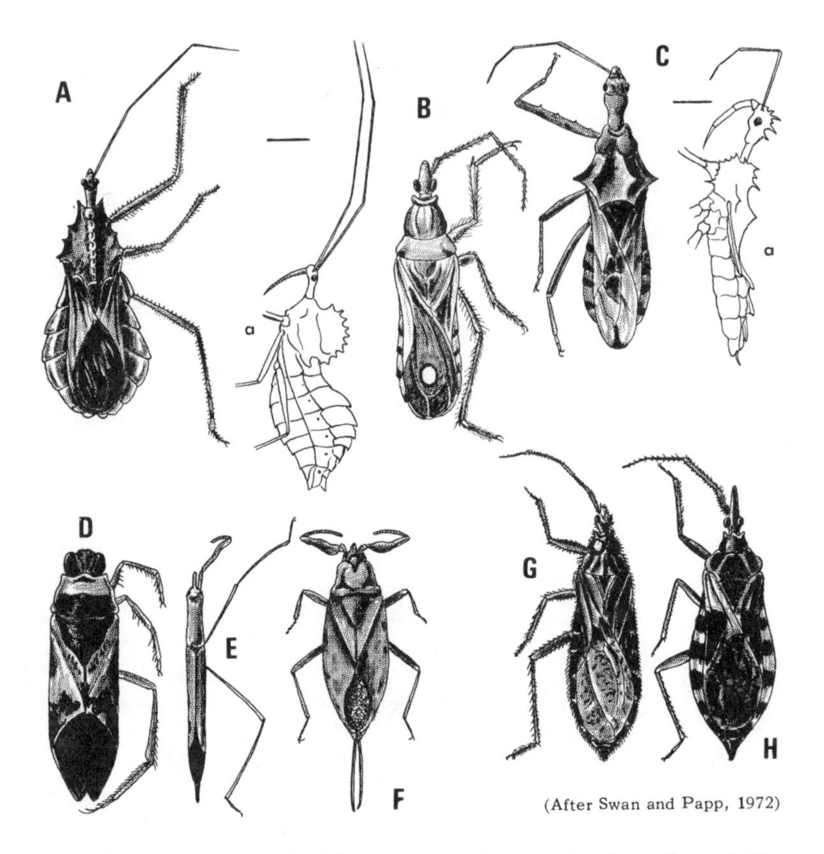

Fig. 39. True bugs: A — wheel bug *(Arilus cristatus),* (a - lateral view); B — western corsair *(Rasahus thoracicus),* a reduviid bug capable of inflicting a painful bite; C — spined assassin bug *(Sinea diadema),* a predaceous reduviid bug, shown for comparison (a - lateral view); D — backswimmer *(Notonecta insulata);* E — western waterscorpion *(Ranatra brevicollis);* F — flat water-scorpion *(Nepa* sp.); G — masked hunter *(Reduvius personatus);* H — blood-sucking conenose *(Triatoma sanguisuga).*

adults of various insects. Another reduviid, called the **Wheel Bug** *(Arilus cristatus)* because of the crest resembling a wheel on its thorax, is a voracious predator and will bite humans readily. A large brown bug, about 28 mm long, it attacks large caterpillars such as hornworms, sucking them dry; it is common in the eastern and southern states, ranges westward into New Mexico, and has been reported in Nevada but not in California. The bites of these predaceous bugs are only in self-defense.

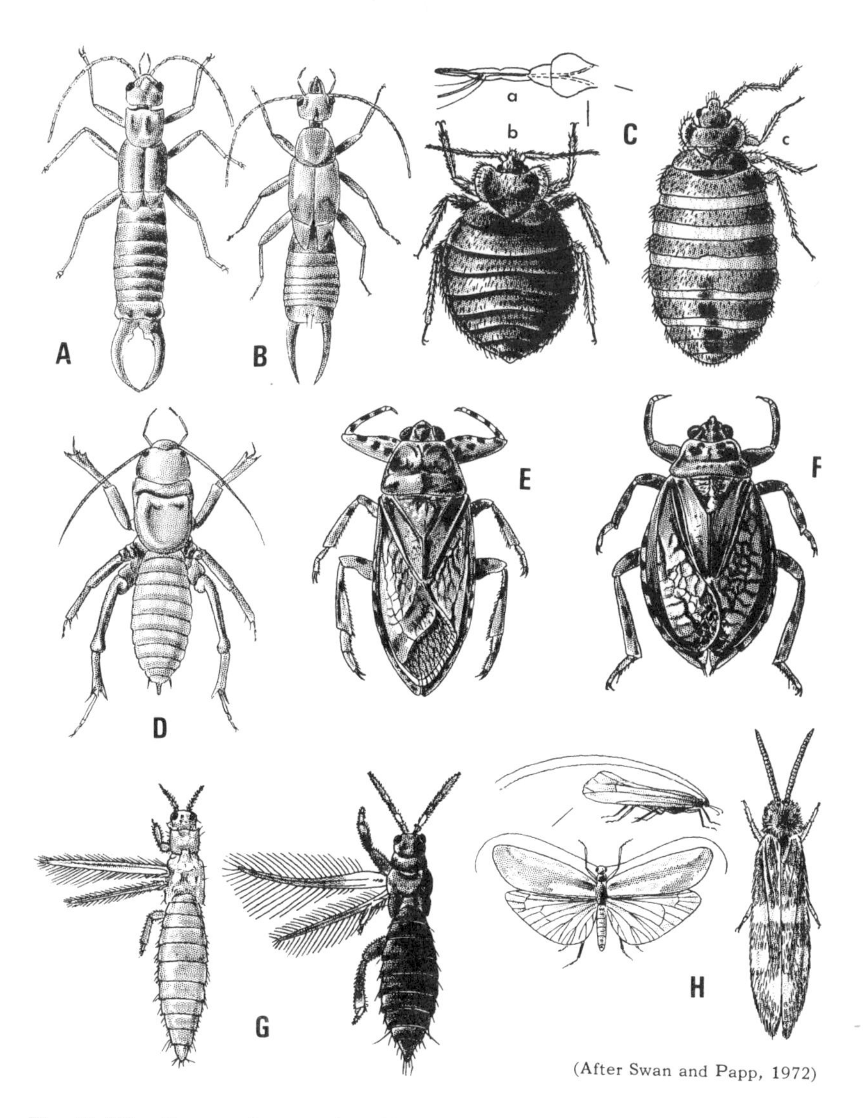

(After Swan and Papp, 1972)

Fig. 40. Miscellaneous insects: A — European earwig *(Forficula auricularia);* B — ring-legged earwig *(Euborellia annulipes);* C — bed bug *(Cimex lectularius)* and details of mouthparts (a), normal (b), and engorged bug (c); D — Jerusalem cricket *(Stenopelmatus fuscus);* E — giant water bug *(Lethocerus americanus);* F — toe biter *(Abedus identatus);* G — trips, and H — caddiceflies, are typical of their groups (Thysanoptera and Trichoptera, respectively).

Water Bugs

Some water bugs are severe biters; as in the case of other predaceous bugs, their bites are defensive. The families dealt with here — the giant water bugs (Belostomatidae), waterscorpions (Nepidae) and backswimmers (Notonectidae) — "are aquatic in all stages and leave the water as adults only during certain flight or dispersal periods" (12), at which times they are attracted to lights, and sometimes find their way into swimming pools. The giant water bugs include some of our largest insects and have been called "fish killers," "toe biters" and "electric light bugs"; bites are usually inflicted when they are stepped on with bare feet. They live in shallow water, are brownish in color, broad and flat, and have a stout beak. When removed from the water, they feign death and eject a fluid from the anus. Some females *(Belostoma* and *Abedus)* glue their eggs to the backs of the males, and the latter carry them about until hatched. The **Giant Water Bug** *(Lethocerus americanus)* is widely distributed throughout the United States, Canada and Mexico, and is one of the commonest species in California; it is medium brown, 40 to 61 mm long and 16 to 24 mm wide, middle and hind femora with three dark brown transverse stripes. *L. angustipes* ranges from Mexico into California; it is medium brown, 47 to 68 mm long, may be distinguished from *americanus* by the broad dark brown streak on the upper side of the front femur, and the absence of the transverse stripes on the middle and hind femora.

Members of the genus *Belostoma* and *Abedus* are considerably smaller than those of *Lethocerus,* more oval and flatter. *Belostoma flumineum* is an eastern species but also occurs in Colorado and California; it is yellowish to blackish brown, 20 to 30 mm long, with three spots on both the inner and outer sides of the front femora. *B. bakeri* is widespread in distribution and common in California; it is about the same size as *flumineum,* may be distinguished from it by the conspicuous gray pubescence along the inner margin of both eyes, and the absence of spots on the front femora. The **Toe Biter** *(Abedus indentatus)* occurs in Mexico and California, and is a fierce biter; it is brown, about 30 mm long, distinctly oval in shape. The giant water bugs described here are those listed as California species by Usinger (96).

Waterscorpions are ferocious predators and will bite readily if handled. They may be recognized by the long tail-like respiratory organ protruding from the anal end of the body, and the scissors-like claws on the front legs. The respiratory organ consists of two appendages which, when held together, form a tube. Some waterscorpions are long and slender *(Ranatra),* others are broad and flat *(Nepa);* the genus *Nepa* is confined to the eastern United States and Canada. The **Western Waterscorpion** *(Ranatra brevicollis)* is the dominant species of this group in California and confined to this state; it is dark brown, the body about 25 mm long. *R. fusca* extends from eastern Canada and United States westward to Colorado and northern California and southward to Mexico; it is dark reddish yellow to brownish black, the body 35 to 42 mm long, the breathing tube 22 mm long. One other species *R. quadridentata,* occurs in California, along the Colorado River; its range extends to Mexico, Arizona, Colorado and eastward. It is uniformly dark brown, body 25 mm long, respiratory appendage same length as body.

The backswimmers are also fierce predators and may inflict a painful bite if handled or stepped on barefooted. They are attracted to lights and thus frequently found in swimming pools, often flying long distances, sometimes in large swarms. The notonectids are boat-shaped, the back convex, which keeps them on an even keel as they swim about on their backs. The long hind legs, fringed with hairs, are adapted for swimming and used like oars. *Notonecta* and *Buenoa* exhibit a striking difference in swimming ability; the former swim with jerky movements, while the latter swim gracefully and seemingly with greater ease. *Notonecta undulata* is widely distributed in North America and occurs in most of the western states, varies in color from dull greenish yellow to almost black (western specimens are usually pale), the hemelytra white, with black wavy transverse band across base of wing membrane and distal end of hardened area (the corium); it is 10 to 12 mm long. *N. unifasciata* occurs from Mexico to Western United States and Canada, is typically black and white, the hemelytra white, with wide black band across base of membrane and distal end of corium; it is about 10 mm long. *N. kirbyi* is widely distributed in the West, typically tan and black, or red and black, with black marking on membrane and corium irregular and extensive; it is about 15 mm long.

Buenoa margaritacea is the most widely distributed species of this genus, ranging from Manitoba to Mexico, and New York to California; it is whitish to dark brown in general coloring, head and thorax solid white to brownish yellow, scutellum yellowish to brownish yellow, sometimes with two anterolateral black spots, females 6 to 8 mm long, males 6 to 7 mm long. *B. scimitra,* widespread in the southern United States from Florida to California, closely resembles *margaritacea,* is white to dark brown in general coloring, head and thorax solid white to brownish yellow, scutellum orange to reddish yellow, sometimes with anterolateral portions black; it is smaller than *margaritacea,* females 6 to 7 mm long, males 5 to 6 mm long. *B. uhleri* occurs in Mexico, Texas and southern California, is usually larger and more robust than *margaritacea,* whitish to gray in general coloring, head and thorax whitish to pale brownish yellow, scutellum usually orange, with irregular black areas at base, 6 to 8 mm.

Bed Bugs

Bed bugs (family Cimicidae) are by no means the prerogative of slums and the slovenly; anyone is liable to have them brought into the house by one means or other, or to encounter them in the uncertainties of travel. They may be transported in luggage, boxes or furniture. Their characteristic odor and shape make their presence and identification unmistakeable. The odor comes from an oily secretion which exudes from glands on the underside of the thorax. The **Bed Bug** *(Cimex lectularius)* is brownish to purplish, 4 to 5 mm long, round and flat, enabling it to crawl through narrow cracks — until it has become engorged. It is flightless, having only vestigial wings which barely extend over the basal segment of the abdomen; it attacks man, chickens, mice, rats and rabbits. The **Bed Bug** is world-wide in distribution; it is primarily an insect of temperate climates but has spread into tropical regions, where the closely related and even more disagreeable **Tropical Bed Bug** *(Cimex hemipterus)* occurs.

Nature of Bug Bites

The defensive bites of predaceous bugs, such as the **Western Corsair** and the water bugs, are painful but do not normally involve any complications. The bloodsucking conenose usually makes several punctures when feeding and the victim feels no pain. The bug takes from eight to ten minutes to engorge itself; reports invariably indicate the sleeping host was unaware of being bitten, until later. As with other bites, the reaction varies with the amount of toxic venom injected and the sensitivity of the person; this ranges from only a slight swelling and redness in many cases to severe eruptions of the skin, and in extreme cases may be preceded by nausea, rapid breathing and pulse and palpitations of the heart. Some species of *Triatoma* found in Mexico, Central and South America transmit the organisms *(Trypanosoma cruzi)* causing Chagas' disease, which is endemic in some areas. Wood and associates found many specimens of the **Western Bloodsucking Conenose** in various parts of California which were naturally infected with these organisms (104, 105).

The bites of the **Bed Bug** have a characteristic 1-2-3 linear pattern and cause itching, red spots and sometimes swellings in sensitive persons. The reaction is believed to be caused by salivary secretions injected when the bug inserts its beak, as well as from the wound itself. As in the case of the bloodsucking conenose, it is suspected that these secretions have anesthetic properties since the bug can feed on sleeping persons three to five minutes — the time it takes to become engorged — without waking the victim. The rude awakening comes later with the itching. According to Smart, "There is no satisfactory evidence that (bed bugs) are the normal vectors of any (diseases) under natural conditions" (82).

Beetles

Beetles comprise by far the largest group (order Coleoptera) of insects measured by the number of species known, but not many of them are troublesome as biters or known to have medical importance (89, p. 330-507). They are easily recognized by the hard wing covers, called elytra, which are the forewings and fold protectively over the membranous hindwings and body when at rest. Some beetles are unable to fly because the elytra are fused at the middle (suture) line of the fold, and the hindwings are missing. Blister beetles are the most important of these insects medically since they contain cantharidin, a vesicant which causes blistering when the insects are crushed on the skin. The agent has been widely used in the past as a counter-irritant, diuretic and aphrodisiac; its use as a drug is dangerous and generally discredited today.

Tiger Beetles

Some tiger beetles (family Cicindelidae), which are fierce predators, are known to inflict painful bites when handled. Those of the genus *Omus* — a small group confined to the Pacific coast of North America — are well known for this trait. The **California Black Tiger Beetle** *(Omus californicus)* is found under stones in dry places, from southwestern British Columbia to the Los Angeles area of California; according to Wallis, this species is rare (98). **LeConte's Black Tiger Beetle** *(O. lecontei)* — largest of the three — is black to blackish brown, occurs in rotten stumps, from British Columbia to western

Fig. 41. A tiger beetle *(Cicindela oregona)* in its natural pose.

Montana and California; it may be distinguished from the others by the deep irregular pits on the elytra. These beetles are flightless since the elytra are fused at the sutura; they are from 12 to 17 mm long, move swiftly and bite viciously.

The **Oregon Tiger Beetle** *(Cicindela oregona)* — a more abundant and widespread species — is also known to bite. This beautiful beetle is usually olive-green, with a bright metallic luster and yellowish markings; some forms are bright blue. It may be distinguished from other tiger beetles (with one exception, the female of *C. scutellaris)* by the small patch of long hairs near the inner edge of each eye. It is from 12 to 15 mm long, widely distributed in the West, from New Mexico to Alaska; like most species of *Cicindela,* it is a good flier and can run fast.

Charcoal Beetles

The **Charcoal Beetle** *(Melanophila consputa),* one of the flatheaded or metallic wood borers (family Buprestidae), is attracted to the smoke of forest fires and bites severely. The beetles gather around firefighters, biting them as the female beetles try desperately to lay their eggs on the smoldering trees. The beetle is black, usually has twelve small yellow spots on the elytra and is from 8 to 13 mm long. The larvae (to which the "flatheaded" part of the common family name applies) bores into the bark and outer layer of injured or fire scorched pines preferably, in California, Arizona and Oregon. *M. acuminata,* which is about the same size as *consputa* but all black, has similar habits. It occurs throughout most of North America, including the Pacific coast states from California to Alaska, New Mexico, Colorado and Idaho; the larvae attack spruce, Douglas fir, cedar and pines.

Blister Beetles

Blister beetles (family Meloidae) are unusual among beetles in that the adult's food differs from that of the larvae. They comprise a small family, but are very numerous at times in some places. The adults are plant feeders and include some serious crop pests; most of the larvae feed on the egg pods of grasshoppers or are parasitic in the cells of wild bees. Females in the first case lay their eggs in the ground and the larvae search out the grasshopper egg pods; in the other case, the eggs are usually laid on flowering plants, where the young larvae latch onto the flower-visiting bees for a ride to the host's nest. Most famous of the blister beetles is no doubt the **Spanishfly**

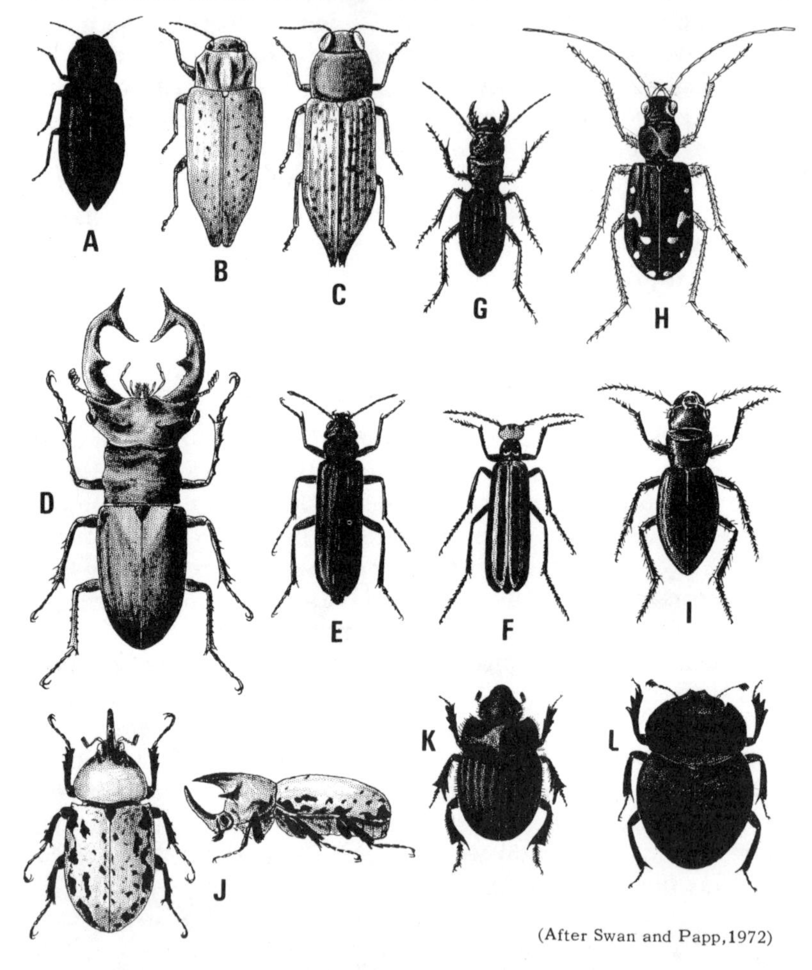

Fig. 42. Beetles: A — charcoal beetle *(Melanophila acuminata);* B and C — flatheaded borers, more typical of the family (Buprestidae) than A, are shown here for comparison; D — giant stag beetle *(Lucanus elaphus);* E-F — blister beetles *(Lytta nuttalli* and *Epicauta vittata,* respectively); G — LeConte's black tiger beetle *(Omus lecontei);* H — Oregon tiger beetle *(Cicindela oregona)* is more typical of the tiger beetles (Cicindelidae); I — California black tiger beetle *(Omus californicus);* J — rhinocerus beetle *(Dynastes tityrus);* K-L — *Copris carolinus* and *Canthon laevis* respectively, are typical of the dung beetles *(Scarabaeidae).*

(Lytta vesicatoria), is bright green, known from central to southwestern Europe, where it feeds chiefly on privet, lilac and ash. The adults of *Lytta, Meloe* and other blister beetles have had a long history of medical use. Pulverized blister beetles were recommended for relief of many ailments as recently as in the 1890's. They were used in many parts of Europe as a treatment for hydrophobia; urination of blood — a symptom of cantharidin poisoning and kidney damage — was interpreted as evidence of a cure. External applications of crushed beetles were used in the treatment of rheumatism in Sweden (71). The irritant — a clear yellowish fluid — exudes from leg joints (femora and tibiae) when the beetles are alarmed and serves as a defensive mechanism. Slight pressure on the beetles will cause the fluid to exude in sufficient amount to produce blisters on the skin.

The **Nuttall Blister Beetle** *(Lytta nuttalli),* a large metallic green or purplish species with sparse short pubescence on the elytra, and black antennae, from 16 to 28 mm long, occurs in the Rocky Mt. region from New Mexico and Colorado to Wyoming and Montana; the adults feed on leguminous plants, the larvae on grasshopper egg pods. The **Green Blister Beetle** *(Lytta cyanipennis)* is green or purplish, from 13 to 18 mm long, has the same feeding habits as *nuttalli;* it ranges from California north to British Columbia, east to Utah, Wyoming and Montana. The **Infernal Blister Beetle** *(Lytta stygica)* is bright metallic green to bluish black, from 9 to 14 mm long; it occurs from California to Oregon and Washington, feeds on various wild flowers and shrubs, also ornamentals.

The **Spotted Blister Beetle** *(Epicauta maculata)* is black, with dense grayish pubescence excepting small spots where the black shows through, from 10 to 15 mm long; it extends from California to British Columbia, New Mexico and Colorado to Montana and Idaho, where it is abundant and injurious, being found on clover, alfalfa, beans, beets, potatoes and other field and garden crops. The **Black Blister Beetle** *(Epicauta pennsylvanica)* — sometimes called the "black aster bug" — is all black, with sparse pubescence, from 7 to 13 mm long, occurs over most of the United States, from New Mexico and Colorado to Montana in the western states; it is found on a wide variety of ornamentals, especially asters, and on goldenrod and other wild flowers. The **White Spotted Blister Beetle** *(Epicauta pardalis)* is black, with numerous spots and white lines, extends from Mexico northward into New Mexico, Arizona, California and Oregon; it is about 10 mm long, usually abundant on grasses and weeds but often invades fields of cultivated crops, more especially potatoes and corn.

The **Giant Soldier Beetle** *(Tegrodera erosa)* and a closely related species *(T. latecincta)* are rather unusual blister beetles, the head and prothorax being red, remainder of body black, elytra yellow, with network of lines, black margins and black band across the middle; they are from 17 to 30 mm long, and occur, often in great numbers, in arid regions of southern California below 4,000 ft.; food plants are sagebrush and alfalfa, the latter often being damaged severely. *T. latecincta* may be distinguished by the darker head and prothorax and more pronounced black markings on the elytra.

Nature of Blister Beetle Injury

Cantharidin — the vesicant or blistering agent possessed by blister beetles — is readily absorbed by the skin. The oily substance is described as "the lactone of cantharidic acid." The first reaction after contact with the fluid is usually just a slight burning sensation. A few hours after exposure large blisters form at the points of contact with the skin; this contact alone may be sufficient to cause irritation of the kidneys. In many cases contact with the beetles occurs at night during outdoor activities under lights, to which the beetles are attracted. In merely brushing the beetles off, enough of the oil may get on the skin to cause blistering.

Stinging Beetle-Larvae

The larvae of dermestid beetles (family Dermestidae), which feed on dead animals and plant materials, sometimes have sharp hairs which are irritating to sensitive persons when coming in contact with the skin or swallowed in food. Several species of *Trogoderma,* which are closely related to carpet beetles and notorious pests of stored foods, are the principal offenders. One of them, most often found in packaged food products in California, is the **Warehouse Beetle** *(Trogoderma variabile).* The adult is brownish black, about 3 mm long, and a good flier. The larva varies from yellowish white to dark brown (as it gets older), and is about 6 mm long. They commonly feed together on stored foods. A world-wide species, it is found throughout most of the United States (68).

Nature of Beetle-Larvae Stings

The larva of the **Warehouse Beetle** (and its close relatives) has numerous setae, which are of two kinds; short one, shaped like a spear and barbed, encircling the abdomen in rows; long straight ones having short pointed hairs and occuring in tufts along each side. These setae cause irritation when coming in contact with the skin of sensitive persons, and may cause digestive disturbances when ingested with food containing the larvae or its hairs. Most digestive disorders of this kind — a condition called canthariasis — appear to occur in children, especially infants, and have usually been traced to packaged cereals prepared for babies. *Trogoderma glabrum,* a species fairly common in the United States, was involved in a case in Indiana, *T. ornatum* was implicated in a case in California (67). Suspected cases should be referred promptly to a physician.

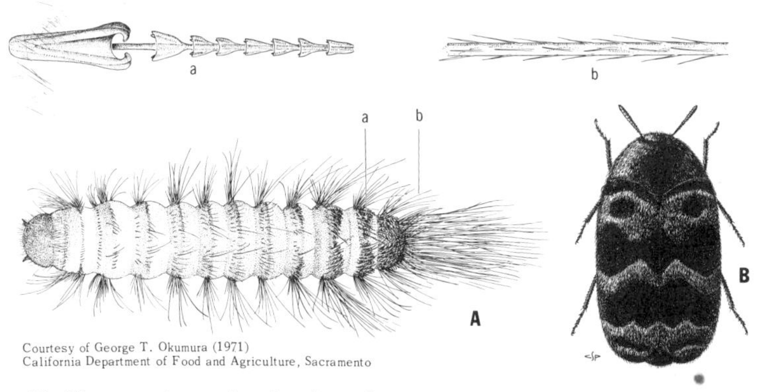

Fig. 43. The warehouse beetle: A — larva or young, showing two types of stinging hairs, hastisetae (a), and spicisetae (b); B — the adult beetle.

Darkling Beetles

Some darkling beetles (family Tenebrionidae) when threatened give off a foul-smelling defensive substance which can be irritating to persons handling them. Some darkling beetles, especially in the genus *Eleodes,* have the amusing habit of standing on their heads when threatened. The posture enables them to direct the defensive spray at their enemies with telling effect. The beetles are pitch black, without wings (the elytra are

Fig. 44. — Darkling beetles, known as "stink bugs." A — *Cryptoglossa laevis;* B — *Eleodes armatus* in protective position; C — Same, playing dead.

fused), and slow in movement. The beetles feed on foliage, the larvae on fungi. The defensive substance, secreted in a pair of glands lying near the tip of the abdomen on either side of the anus, has been found to contain chemicals known as quinones; it is very irritating to the eyes, and causes a painful reaction if it comes in contact with cuts or scratches on the skin. If many of the beetles are handled the fingers become stained.

A very common species in southern California and Baja California, which readily assumes the headstand posture, is *Elodes acuticaudus;* it is shiny and smooth, pitch black, elongate, with elytra coming to a point at the posterior end, about 25 mm long or less on the average. It is often found in association with *Coelocnemis magna,* which resembles it closely and has the same habits; *C. magna* is larger, about 30 mm or more long, pitch black, with sides more rounded, posterior end tapered but not coming to a sharp point as in the case of *E. acuticaudus.* They are nocturnal but frequently seen during the day, usually walking slowly but capable of travelling at a fairly good pace.

VII. STINGING CATERPILLARS

Caterpillars are the larvae of butterflies and moths, which comprise a group (order) called Lepidoptera. The name means "scaly wings," and is descriptive of the overlapping scales that cover the two pairs of wings, and the body and legs of the adults. Butterflies may be distinguished from moths by the antennae, which are clubbed in the butterflies and tapered and hairlike or feathery in the moths; the feathery antennae are an adornment of the males. Another distinction is in the way the forewing and hindwing are held together in flight. Most moths have a "hook and eye" mechanism consisting of one or more bristles projecting from the front margin at the base of the hindwing, which hooks into a flap (in the males) or a row of bristles (in the females) on the underside of the forewing. Butterflies have a simple lobe on the hindwing which grips the underside of the forewing. The majority of moths rest with the wings spread out and to the sides, or fold them over the body; butterflies generally hold them upright when resting. There are exceptions in either case. Most butterflies are active in the daylight hours, while moths are generally active at night or in the twilight hours, with some exceptions.

Caterpillars can usually be distinguished from the larvae of other kinds of insects by the prolegs. These are the short, fleshy projections on the underside of the abdomen which are used in walking and clinging to plants, and serve as an adjunct to the three pairs of true legs attached to the thorax; the prolegs completely disappear when the larva pupates and changes into the adult form. Caterpillar prolegs vary in number from two or three pairs (in the "loopers") to five pairs, including the pair on the hind end, and have a ring of crochets or hooks on the bottom. The larvae of flannel moths (family Megalopygidae) which have seven pairs of prolegs are an exception. The larvae of sawflies (order Hymenoptera) resemble caterpillars but they have six to eight pairs of prolegs and lack crochets.

"Stinging caterpillars" bear hairs or spines having urticating properties, that is, when coming in contact with the human skin they act like nettles, causing a stinging and itching sensation. This may be followed by a skin eruption and sometimes a systemic reaction, varying in intensity with the species of caterpillar and the sensitivity of the person involved. There are said to be in all about 50 species of caterpillars (in 10 families) having stinging hairs (40). They occur largely among the moths, not often among the butterflies.

Giant Silkworm Moths

The giant silkworm moths (family Saturniidae) are noted for their colorful markings and large size. They are called "silkworm" moths because of the dense silken cocoons which some of the larvae spin before pupating and changing to the adult form. These are not the same silkworms (family Bombycidae) that are used for the production of silk thread. Some saturniid caterpillars have stinging hairs, notably the **Io Moth** *(Automeris io).* The caterpillar of this moth is pale green with pink and white lateral stripes, about 38 mm long full grown; it has numerous clusters of green spines (with a few black ones) radiating from tubercles. Some of the long spines have hairs on the tips, but the poisonous ones have peg-like tips connected to venom glands and break off in the skin, causing intense itching (they do not float in the air as in the case of the **Range Caterpillar** described below). The **Io Moth** occurs in the eastern and central states, west to Colorado, New Mexico, Texas (Guadalupe Mts.). The larva feeds on a wide variety of trees and shrubs, also corn. The adult male moth is bright yellow, the female reddish brown and larger, wing spread about 75 mm; the hindwing has an orange band near the margin, a black wavy band and black "eye spot" near the middle.

Probably most, if not all of the species in the genus *Hemileuca,* have stinging hairs. Comstock and Dammers state that all of those with which they were familiar, "produce a stinging sensation when touched, which may be followed by a rash." Unlike *Automeris* and others, the stinging hairs are not lodged in the pupae, which are often thin. Practically all of the moths in this genus are day-flying. The caterpillar of the **Buck Moth** *(Hemileuca maia)* is dark brown with yellowish spots along the

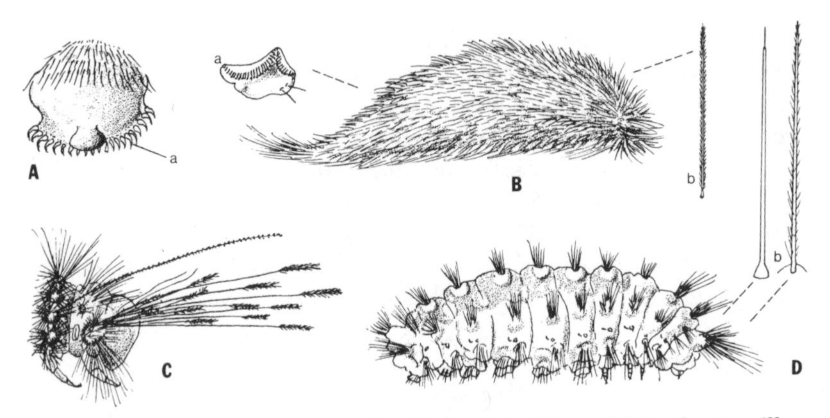

Fig. 45. Stinging caterpillars: A — detail of proleg of the saddleback caterpillar (see also Fig. 50) showing crochets (a) which are used for clinging to surface; B — pus caterpillar, showing detail of proleg and crochets (a) and venemous hair (b); C — head of white-marked tussock moth caterpillar showing long plumose setal tufts projecting forward; D — hackberry leaf slug caterpillar, side view, of a flannel moth *(Norape ovina),* showing details of venemous hairs (b).

back and sides, and branched stinging hairs rising from tubercles; it is about 38 mm long full grown, feeds on oak and willow. Strawberry pickers in Maryland, according to a USDA report, were forced to stop picking because of stings of this caterpillar. The adult **Buck Moth** *(Hemileuca maia)* — named for its habit of flying when the deer is hunted — is reddish brown, with broad white medial band across both wings, abdomen with tuft of bright red hairs on tip, wing spread from 50 to 75 mm; as in other members of the genus, the male antennae are bipectinate. An eastern species, it ranges west to Kansas and Texas. The western form, the **Nevada Buck Moth** *(H. nevadensis)* is similar, the larva greenish with brown spots along the back and sides; in the adult the light portion of the wings is much more extensive. A day-flying moth, it appears in the fall, occurs in Texas, New Mexico, north to Colorado, Nebraska, the Dakotas, Minnesota, Wisconsin, Manitoba, Saskatchewan, west to Arizona, California, Utah and Oregon.

The larvae of **Burn's Buck Moth** *(Hemileuca burnsi)* is black, with numerous small round white dots with black centers and soft white hairs, eight rows (four on a side) of long black multiple-branched spines, each with grayish white tip, spiracles orange with black rims, arching yellowish bars below the spiracular line, about 50 mm long full grown; it is found in "the sagebrush country on the slopes of the desert mountains" of

Arizona, Nevada, Utah and southern California, its favored food cotton thorn *(Tetradymia)*. The adult is creamy white with black markings. The larva of *H. electra* is blackish, with white dots, thin black middorsal line, two white subdorsal lines, orange spiracles and the usual branching spines, which are all black except the yellow tips in the lower rows, about 45 mm long full grown; it occurs in southern California and western Arizona, feeds on wild buckwheat. The larvae feed gregariously in the early stages, as do those of other species of *Hemileuca*. In the adult the body is orange with black spots laterally and sometimes dorsally, the forewing marked with black and whitish or yellowish, hindwing reddish orange with black border and black median spot, spread 50 to 55 mm. The larva of *H. juno* is very similar to that of *electra,* differs in the color of the branching spines, which are rose-pink at their bases and black at the ends; it is 45 to 50 mm long full grown, feeds on cottonwoods, willows and poplars, ranges from northern Mexico to New Mexico and Arizona, north to Idaho, west to the arid regions of California. The adult moth may be recognized by the oblique white band across the forewing and the solid black hindwing.

The **Range Caterpillar** *(Hemileuca oliviae)* causes asthmatic and acute caterrhal attacks among cattlemen in New Mexico. It is a serious pest of range grasses, extends from New Mexico north to Colorado, east to Kansas, Oklahoma and Texas, south to Mexico. Caterpillar populations increase to injurious levels in 10- to 12-year cycles. The caterpillar is yellowish, gray or black, with dark red or black head, white spiracles ringed with black and dense coat of coarse spines with stinging properties, about 60 mm long full grown; the stinging hairs are very irritating to the skin and tend to float in the air, causing respiratory distress when breathed in. Besides wild grasses, the larvae not infrequently feed on cultivated crops. The adult has a reddish brown to black body and buff or clay-colored wings; the forewing has two wide, lighter colored lateral stripes, the hindwing is plain.

The larva of the **Sheep Moth** or **Brown Day Moth** *(Pseudohazis eglanterina),* as in the closely related genus *Hemileuca* [Ferguson changed *Pseudohazis* to *Hemileuca,* with some hesitancy (30).], has spines with stinging properties. Packard wrote: "The dorsal spines are shorter and sharper than those of *H. maia,* being very sharp and the prick painful even in alcoholic

Fig. 46. Stinging caterpillars: A — mourning cloak butterfly caterpillar *(Nymphalis antiopa);* B — white-marked tussock moth caterpillar *(Orgyia [Hemerocampa] leucostigma);* C — brown-tail moth caterpillar *(Nygmia phaeorrhoea).*

specimens." The caterpillar is dark brown to black, with reddish spots on the back and a narrow red line on each side, stout black and tan branched spines, from 50 to 60 mm long full grown; it feeds on manzanita, ceanothus, wild blackberry, cherry and grapes, cultivated grapes, cherries, plums, prunes and many other plants. Its range is from British Columbia to California, east to Colorado. The adult is yellowish, with pink shading, curving black lateral band across both wings, black shading along outer margins, a large black circular patch near center of each wing. The adult of the **Sagebrush Sheep Moth**

Fig. 47. Anterior region enlarged. Caterpillars shown in Fig. 46.

(P. hera) resembles that of *eglanterina* closely; the patch near the center of the wings is larger and more oblong, the whole basal area or more of hindwing black, spread from 52 to 69 mm. It occurs in the sagebrush areas of the Rocky Mts. and foothills,

and in the Great Basin: Montana, Wyoming, Colorado, west to the Sierra Nevada Mts. of California, north to the interior of southern British Columbia. The caterpillar has black spines and spinules with branching whitish hairs, feeds on sagebrush.

The larvae of *Saturnia mendocino* is uniformly pinkish yellow, with numerous spines having stinging properties; the head is chocolate-brown, with numerous white setae, the spiracles are orange with reddish brown rims and light yellow line below from anal end to thorax. It is about 30 mm long full grown, occurs in the central and northern parts of California; its favored food is Manzanita. Comstock said of this species: "Care must be observed in handling the larvae as the spines can sting the hands as do those of *Pseudohazis.*" The male moth is reddish brown above, hindwing bright yellow except dark base and black lateral band near outer margin; both wings with yellow and blue and black eyespot, spread about 62 mm.

Tussock Moths

Tussock moths (family Lymantridae) are named for the tussocks or tufts of hair on the caterpillars. The tufts in some species contain stinging hairs, the short barbed ones causing intense skin irritation. The larva of the **Whitemarked Tussock Moth** *(Hemerocampa leucostigma)** is one of the principal offenders in this group. It is light brown, with a velvety black band on the back and two yellow stripes below this on each side; the head is vermillion as is a spot on each side of the sixth and seventh abdominal segments. A tuft of long black hairs is located on each side at the base of the head, and one at anal end of the abdomen; four tufts of short white hairs are located in between, on the first four abdominal segments. The stinging hairs are scattered over the body in the early stages and concentrated in the short white tufts in the late stages. The mature larva is from 30 to 35 mm long, the male much smaller than the female; it feeds on deciduous trees and shrubs. The male moth has gray wings with darker wavy lines and spread of 32 mm; the female is wingless. It ranges from eastern United States and Canada to Colorado and British Columbia. The **Western Tussock Moth** *(H. vetusta),* the larva of which is also

*Now *Orgyia leucostigma.*

noted for its stinging hairs, is very similar to the **Whitemarked Tussock Moth,** takes its place in the Pacific coastal region from California to British Columbia. The caterpillar is gray, with numerous red, blue and yellow spots, four white tufts medially and one posteriorly, and two long black pencils anteriorly and one posteriorly. It is from 13 to 25 mm long, feeds on oak, poplar, willow and other trees. The winged male is brown, with gray markings, spread from 20 to 25 mm; the wingless female is gray, 12 to 25 mm long.

Tiger Moths

The full grown larvae of tiger moths (family Arctiidae) are generally covered with a dense coat of tufted hairs arising from large tubercles or warts; in some species the tufts have barbed urticating hairs which are very irritating to sensitive persons. One of the most offensive of this group is the caterpillar of the **Silverspotted Tiger Moth** *(Halysidota argentata),* which is densely covered with tufts of brown and black hairs of varying lengths. The young larvae are gregarious and feed in clusters under webs, where they also hibernate. The older larvae feed singly and are about 38 mm long full grown; host trees are Douglas fir, firs, spruce, hemlock and pine. The adult has a creamy white abdomen, brownish forewing with numerous silvery spots, silvery white hindwing with brown patch near center and at apex, spread from 45 to 50 mm. It is common in the coniferous forests from Colorado to California, Oregon and southwestern British Columbia. In attempting to study this insect, G.T. Silver of the Forest Biology Laboratory in Victoria, British Columbia had great difficulty trying to rear it, on account of the stinging properties of the caterpillar. He reported: "The barbed setae break off easily and the tiny tips become embedded in the fingers and any part of the body touched by the hands, producing red swellings which are very irritating. Mature larvae and cocoons are particularly notorious for this. Most of the insectary staff were allergic and consequently only a few people could be used on *Halysidota* work." (81).

The larva of the **Hickory Tussock Moth** *(Halysidota caryae)* also has a reputation for stinging. It is whitish with black head and middorsal line, has black warts with white tufts, row of eight black tufts along the back, two long black pencils on the fourth and tenth segments, four thin white ones on the second and third segments and two on the eleventh and twelfth segments; it is about 35 mm long full grown. They feed in colonies until after the last molt, when they scatter; they do not make webs. Its range is from Quebec to Virginia, west to Colorado, Texas and California.

The caterpillar of the **Garden Tiger Moth** *(Arctia caja)* is also known for its stinging properties. The body of the caterpillar is blackish brown, with tufts of long black hairs dorsally, yellowish brown on sides, all yellowish brown on the first and fourth segments, spiracles yellowish white; it is 45 mm long full grown. Host plants are plantain, dandelion, nettles, lettuce, poplar, walnut and apple. The adult is highly variable, thorax brown edged with white, abdomen orange with black dorsal spots, forewing brown with broad white bands, hindwing orange with black spots, spread from 45 to 65 mm; it ranges across Canada and the northern states from the Atlantic to the Pacific.

Some Other Stinging Caterpillars

Some leaf skeletonizer moths (family Zygaenidae) have caterpillars with stinging hairs. The larva of the **Western Grapeleaf Skeletonizer** *(Harrisina brillians)* is yellowish with black crossbands, and tufts of long black urticating hairs on each body segment. The larvae are gregarious, feed in compact colonies on underside of leaves, are sometimes damaging to cultivated grapes; wild grapes, Virginia creeper and Boston ivy are also hosts. The adult is metallic blue to greenish black, with long narrow wings, spread about 25 mm. It ranges from Mexico to southern California, Utah, New Mexico, Arizona and Texas. Recently also northern California.

The caterpillars of some brushfooted butterflies (family Nymphalidae) have stinging hairs. One of these is the beautiful **Mourningcloak Butterfly** *(Nymphalis antiopa)* which is widespread in North America and occurs in Europe and Asia. Sometimes called the **Spiny Elm Caterpillar,** the larva is black, with rows of branched spines having urticating properties; it has tiny white dots and a row of red dots on the back. The wings of the butterfly are velvety and purplish brown, with yellowish border edged on the inside with blue spots, spread from 55 to 70 mm. The caterpillar feeds on willow, poplar and elm.

The larvae of some flannel moths (family Megalopygidae) are among the worst of the stinging caterpillars in the effect of their stings on the skin and systemically. They occur mostly in the eastern and southern states. The furry **Pus Caterpillar** *(Megalopyge opercularis),* probably the best known and the most troublesome, ranges from Virginia to Texas. According to Bishopp, "it produces the severest sting of all the forms (of stinging caterpillars) occurring in the United States." Several species in the family are recorded from southern Arizona but they are not well known. Comstock found what he called "the little yellow 'Flannel-moth'," *Dalcerides ingenitus,* in the Santa Rita Mts. of Arizona; he described the larva as "slug-like and yellow." Some slug caterpillars (family Limacodidae or Eucleidae) are also known for their stinging properties, but they are eastern and southern in distribution. A few species have been recorded from southern Arizona and California but they are not well known. The subspecies *Parasa chloris huachuca* has been recorded from southern Arizona; the caterpillar of this species is slug-like and has stinging hairs; the body is oval-elliptical in shape, chestnut-brown dorsally, yellowish below, margin of dorsum elevated into sharp ridge, tubercles with bunched whitish hairs on third to fifth and tenth and twelfth body segments, anal segment with long spine; from 15 to 20 mm long full grown.

The **Gloveria Moth** *(Gloveria medusa)* — family Lasiocampidae — is possessed of hairs that are irritating to the skin. The body of the caterpillar is gray, speckled with black, the upper half clothed in long erect black hairs and short erect brown hairs, the lower half with tufts of dropping buff-gray hairs; it has yellow spiracles with black rims, is about 70 mm long full grown. It is found in the coastal lowlands of southern California; host plants are wild white lilac *(Ceanothus)* and false buckwheat. The adult is a very large moth, with wing expanse of about 100 mm; the thorax and forewing are dark smoky gray, the latter with a small white discal spot; the abdomen and hindwing are somewhat lighter, with a brownish tint.

Fig. 48. Caterpillars of the western grape leaf skeletonizer *(Harrisina brillians).*

Tent Caterpillars

The cocoons of all species of *Malacosoma,* the tent caterpillars (family Lasiocampidae), are dusted with a white powder which is very irritating to sensitive persons. It produces an intense itching and red welts, a reaction similar to that caused by nettles. The reaction is more severe where the skin is folded and when the person is perspiring (86). The **Western Tent Caterpillar** *(Malacosoma californicum)* and its many subspecies occur throughout western North America. The larva is highly variable, typically; reddish orange to brown above, pale brown below, with a pale blue line on each side and a white one down the middle. The larvae live in colonies, construct tents in forks or crotches of trees for shelter, leave trails of silk as they move away from the tent to feed; host trees are willow, oak, ash, madrona and many others. The adult is also quite variable, reddish brown to yellowish or grayish, usually with a pair of oblique white lines across the forewing, spread from 30 to 40 mm.

Fig. 49. Caterpillars of the western grape leaf skeletonizer, showing stinging hairs. The fly is a tachinid parasite ovipositing in a caterpillar. (Photo by O.J. Smith, courtesy of C.P. Clausen.)

Caterpillar Stings

There are two kinds of stinging hairs found on caterpillars: a small hollow hair seated in a single glandular cell in the epidermis, which secretes the venom; and a shorter hollow spine seated in a tubercle and occuring in rows, with underlying hypodermal venom-secreting glands. When the sharply pointed hairs or spines touch the skin, they break off and release the venom. The true nature of the venom is not presently known, except that it is said to be proteinaceous and acid. The reaction to the hairs may be due in part, or more, to the properties of the hairs themselves than to the venom. Urticating hairs retain their stinging properties for a long time, even after the caterpillar is dead. Hairs imbedded in the cast off larval skin (after molting) and in the pupa or cocoon still have their stinging qualities.

In serious cases of caterpillar stings the reaction may involve inflammation of the skin with swelling and blisters and a severe burning pain; extreme symptoms may include nervousness, nausea, cramps, head pains and difficulty in breathing. Ainslie (1) relates the effect of the stinging hairs of the **Range Caterpillar** as follows. "The irritation lasts for an hour or two, especially when the thin skin on the arm or wrist is wounded. The spot puffs up almost at once, turns white and when the swelling subsides, a brown point remains for days or weeks . . . even the tough skin of the finger tips proves vulnerable and a puncture there is generally very painful." Windborne hairs inhaled can cause an asthmatic attack or a severe caterrhal condition with coughing; those floating in water and taken into the mouth can cause inflammation of the mucous membranes. Nodular conjunctivitis is caused by hairs getting into the eyes.

The reaction from handling cocoons of the tent caterpillars is caused by a secretion of the malpighian tubules of the mature larva, which is excreted through the anus and smeared on the inside of the cocoon with its head, forcing the fluid between the

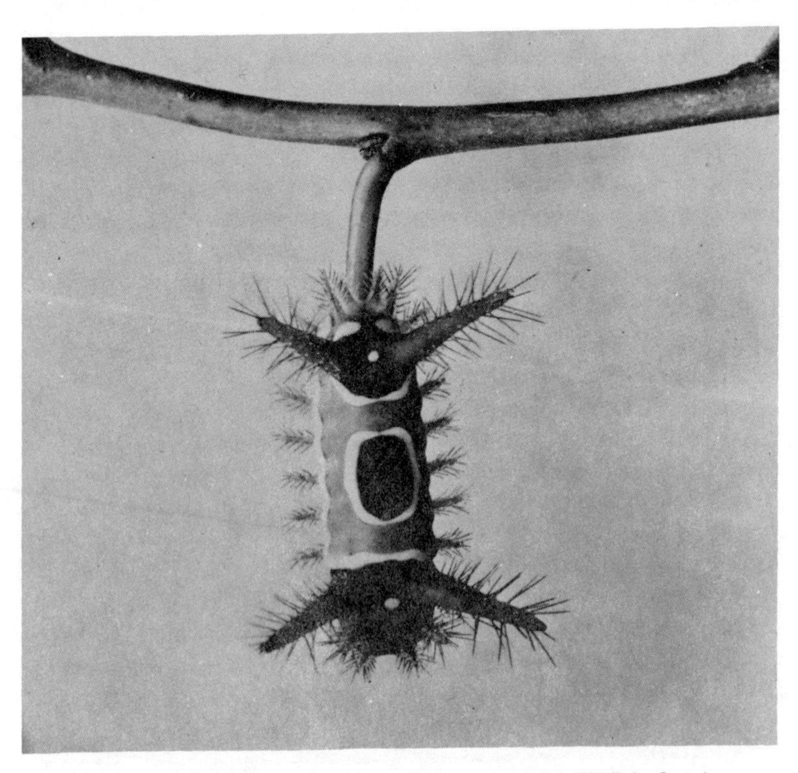

Fig. 50. Saddleback caterpillar *(Sibine stimulea)*, (USDA photo).

silk threads. The liquid dries and cakes and breaks into a fine powder when the cocoon is handled. The reaction is relieved by washing thoroughly with soap and water (86).

First Aid and Prevention

The area affected by caterpillar stings should be washed with soap and water to remove the urticating hairs, spines and venom. Household ammonia or bicarbonate of soda added to the water, or ammonia applied directly, is useful in relieving the itching. Adhesive tape or Scotch tape applied immediately to caterpillar stings will sometimes remove broken spines imbedded in the skin. Ice packs help to relieve the stinging and reduce swelling. In the case of severe reaction, the person should be taken to a physician. It is important that he know the cause of the symptoms and not attribute them to another insect sting or bite. Ointments or lotions applied externally have little value (32).

Avoidance of the caterpillars and cocoons is obviously the best means of prevention. In the home garden the caterpillars can be picked from plants using rubber gloves; when this is impractical, one can use an approved insecticide as directed.

As soon as the green vegetation dries up, rocky areas turn into a haven for scorpions and spiders. Venturing into these areas demands caution.

Normally, the persons most apt to come in contact with stinging caterpillars and cocoons are children (while playing in gardens or trees), campers and forestry workers. Stings most often occur in the fall.

Now you are in tick country. If possible, walk in areas not heavily vegetated. Wear long pants and hiking shoes. Do not lay on the ground, use blankets if you want to rest.

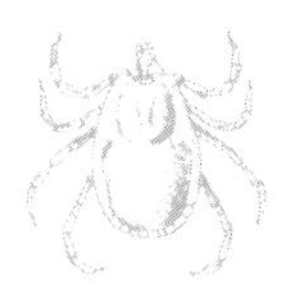

VIII. MITES AND TICKS

Mites and ticks may be free living or parasitic. As a group, they are of great medical importance, many of them being vectors of various kinds of disease organisms affecting wild and domesticated animals and man. Certain ticks also cause "tick paralysis" by injecting toxic substances into their hosts when feeding. Some mites cause various forms of dermatitis in domestic animals and man, others act as intermediate hosts of tapeworms infesting sheep, goats and cattle. Many mites are of great economic significance as pests of cultivated crops and stored food products and drugs, while some mites found on plants are predaceous on plant-feeding mites and on the smaller forms of plant-feeding insects and are, thus, beneficial.

Mites and ticks (order Acarina), as indicated elsewhere (see p. 7), are more closely related to spiders (order Araneida) than to insects. They are easily separated by the fact that typically mites and ticks have only one body segment and are without a true head. To review, insects have three body segments: head, thorax and abdomen; spiders have two body segments: cephalothorax (head and thorax combined) and abdomen. The "head" region of mites and ticks is called the capitulum and projects from the underside or anterior end of the body; it consists of the chelicerae, the "beak" or hypostome and the palps. The chelicerae are paired blades — the cutting organs — which permit insertion of the hypostome in the host's skin; the hypostome is a piercing-sucking organ. The palps are believed to be sensory in function; those of mites are clawed (chelate). Mites and ticks (and spiders), to repeat, do not have the antennae and wings characteristic of insects. They differ also in usually having four pairs of legs in the adult stage and three in the larval stage and in having the legs generally divided into six segments instead of five as in insects. Eyes are usually present, commonly one or more on each side of the body. Mites and ticks usually undergo five or more stages of development, typically: egg, larva, proto-

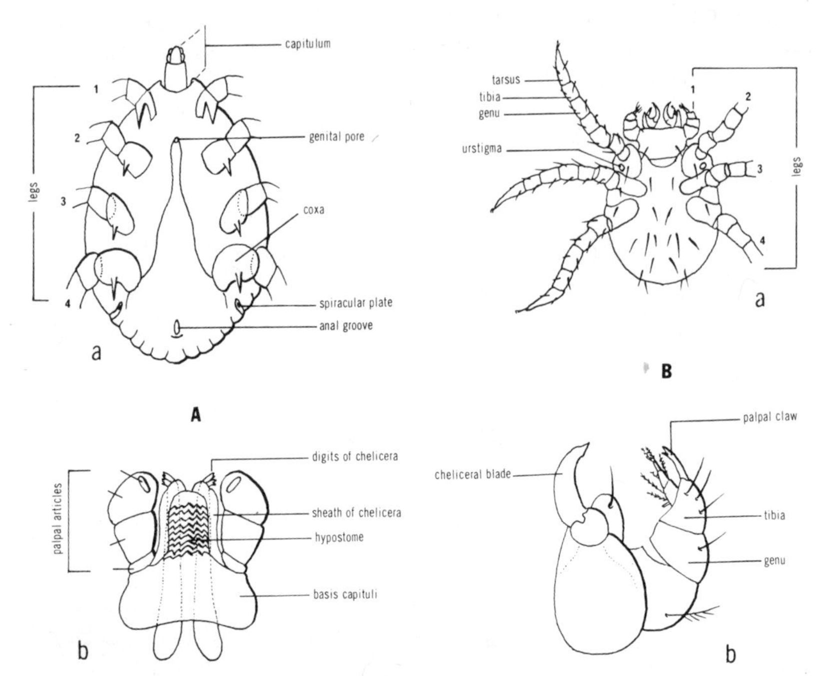

Fig. 51. Tick and mite details: A — underside of male hard tick (a), and details of the capitulum (b); B — underside of the larva of the chigger mite (a) and details of the gnathosoma or "mouth region" (b).

nymph, tritonymph, hypopus and adult. Insects generally have only a nymphal stage or larval and pupal stages in the higher forms, between the egg and adult.

Mites may be distinguished from ticks by their smaller and often microscopic size. Mites are usually less than 3 mm long, while ticks are 3 mm or more in length. Another distinction is in the body covering; in mites it is membranous or protected by hardened areas and often has many long hairs, while in ticks it is leathery and has only a few short hairs. The hypostome — the piercing-sucking organ — is without teeth in the mites and usually has strong, recurved teeth in the ticks; the teeth serve to anchor the tick to the host's skin. A finer point of distinction is in the terminal segment (tarsus) of the first or front pair of legs; ticks have a sensory pore (called Haller's organ) on this segment, mites do not. The parasitic females usually require a blood meal before laying their eggs; some nymphal forms of mites require a blood meal before molting. Some species lay their eggs on the host, others drop to the ground and lay them in the soil. As in the case of some insects, the females of a few species retain their eggs in the body until hatched, giving birth to larvae, or to nymphs or adults.

Mites

Chiggers (family Trombiculidae) are troublesome pests of man, and occur throughout the continental United States. Commonly known as "red bugs," they are found in wooded areas, orchards, berry patches, around lakes and sometimes in lawns, golf courses and parks. One of the most common and widespread species of chiggers attacking man is *Trombicula (Eutrombicula) batatas,* which occurs from South Dakota, Nebraska and Kansas to Pennsylvania and all states southward and in California. In the lower San Joaquin Valley of California, "The bites of *T. batatas* are the cause of extreme annoyance to man" (38); lawns are the usual place from which the mites attack. Only the larval form is parasitic; it is orange-yellow or light red and microscopic in size. The nymph and adult are brilliant red, hairy and very small — about 1 mm or slightly more in length; they feed on small insects, insect eggs and other organisms found in the grass. [The developmental stages of most trombiculid mites are: egg, deutovum (which develops within the egg and contains the maturing larva), larva, nymphochrysalis, nymph, imagochrysalis and adult.] The six-legged larva feeds by inserting the mouthparts in the skin (they do not burrow under the skin); the salivary fluid injected into the host dissolves the tissues, preparing them for ingestion. (Blood is not believed to be an important constituent of the larva's meal; the reddish color is due to pigments in the mite's tissues.) Itching, reddish welts, papules with clear fluid, swelling and sometimes fever, usually develop within 24 hours after attack. Body parts involved are generally the legs, arms and around the waist, and in males often the genitalia, with swelling of the scrotum, penis and foreskin and extreme pain in uncircumcised persons. In California the mite appears from July to October. This species also attacks domestic and wild animals, may be damaging to turkeys and chickens. *Trombicula (Eutrombicula) belkini,* widely distributed in California and also reported from Utah, is close to the eastern *T. alfreddugesi.*

In southern California it attacks man, rodents, ground birds and reptiles (its favored host). In this country chiggers are not known to transmit disease organisms; closely related species in the Far East transmit rickettsiae causing scrub typhus.

The **Chigger** *(Trombicula irritans)* is one of the smallest harvest mites and is closely related to the "red spiders." It is known from New York to Kansas, Minnesota and throughout the South with some closely related species known to occur further to the West. The larvae appear in June (May in the South), resting on various plants in the wild, waiting until a passer-by brushes the leaves. The larva will attach themselves to their new host: humans or animals. The larva do not burrow into the skin, but by attaching themselves to the skin cause intense irritation and reddish spots on the surface of the skin. They commonly attach themselves to humans, rabbits, mice, rats, squirrels; also observed on quail, prairie chickens, toads and turtles. It is also reportedly found on snakes. If humans are attacked by several chiggers, it may result in a high temperature, and cause a nervous disorder to develop, the most common being sleeplessness.

There is another chigger, *Trombicula akamushi,* known throughout Southeastern Asia and the Southwestern Pacific,

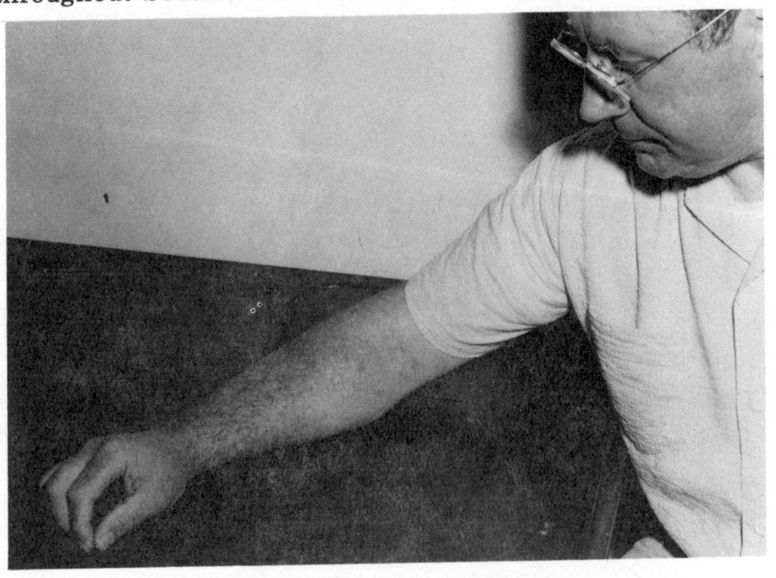

Fig. 52. Chigger bites on man's arm (USDA photo).

which may be picked up and carelessly distributed by unsuspecting travelers. This chigger is credited for the Japanese river fever (also known as tsutsumagushi disease). Infection is about 35 to 40% fatal.

Cheese mites and grain mites (families Acaridae and Glycyphagidae) are nonparasitic but cause mild to severe dermatitis in man. The **Grain Mite** *(Acarus siro)* causes "baker's itch"; the **Grocer's Itch Mite** *(Glycyphagus domesticus)* is the cause

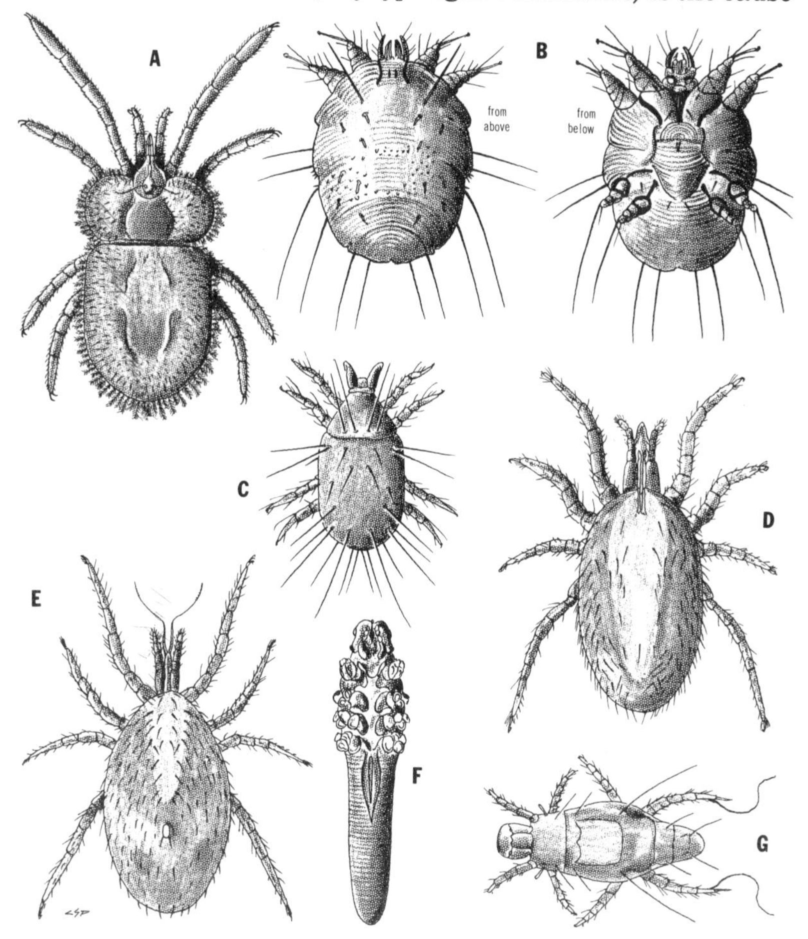

Fig. 53. Mites: A — chiggers *(T. [Eutrombicula] batatas)*; B — itch mite *Sarcoptes scabiei)*; C — a mite *(Glycyphagus* sp.) causing "grocer's itch"; D — tropical rat mite *(Ornithonyssus bacoti)*; E — housemouse mite *(Allodermanyssus sanguineus)*; F — follicle or pore mite *(Demodex folliculorum)*; G — straw itch mite *(Pyemotes ventricosus)*.

of "grocer's itch." The **Cheese Mite** *(Tyrophagus casei)* may be involved in "cheese mite dermatitis." The mites are soft-bodied, slow moving, usually pearly white in color. Whether the irritation caused by handling infested products is due to the dust resulting from feeding by the mites or by the mites themselves, is not known for sure. The reaction of dock workers who handled cheese infested with *Tyrophagus castellanii* is cited by Baker as being possibly typical of the dermatitis caused by other mites of this kind: "About 24 hours after being covered by the powdery debris on the cheeses (composed of mites, skins, feces, etc.), the face and forearms became irritated and the eyelids swelled; there was a diffuse erythema over the face and neck; the forearms also contained small urticarial lesions." It is also stated that, "Besides causing dermatitis, these mites have been found infesting the urinary tract and lungs of man" (3).

Several species of dermanyssid mites (family Dermanyssidae) that attack fowl, rats and mice also attack man, their bites often being the cause of severe itching and painful skin irritation. The **Chicken Mite** *(Dermanyssus gallinae)* is cosmopolitan in distribution, a parasite of chickens and sometimes very damaging to them, also attacks man. Recently engorged females are bright red, specimens with partially digested blood are black or grayish, unfed specimens are whitish; they have a single dorsal plate (narrowing but truncate posteriorly) and long whiplike chelicerae. The adults can survive several months without blood meals. The **House Mouse Mite** *(Allodermanyssus sanguineus),* a parasite of mice and rats, occurs throughout the United States and "is the natural vector of rickettsialpos of man" (90). Cases have occured in New York where the presence of infected mites left little doubt as to their involvement. The color varies from red to blackish or whitish, depending on when the last blood meal was taken. This species is distinguished by the two dorsal plates (the posterior one very small and having two setae) on the adult female; the chelicerae are long and whiplike as in the **Chicken Mite.**

The **Tropical Rat Mite** *(Ornithonyssus bacoti),* a parasite of rodents, is cosmopolitan in distribution and is known to attack persons living in buildings infested with rats. It is an important pest of laboratory animals such as rats, mice and hamsters. As in the case of other blood-ingesting mites, the color varies from red to blackish or whitish. The single dorsal plate is relatively narrow, does not cover the entire dorsal surface; the chelicerae are toothless. The **Northern Fowl Mite** *(Ornithonyssus sylvarium),* found throughout the temperate regions of the world, is a parasite of domestic fowl and wild birds, and will readily attack man in the absence of avian hosts. It is very similar to the **Tropical Rat Mite,** may be distinguished by the wider dorsal plate (which still does not cover the entire dorsal surface) and the position of the setae on the sternal plate, the two anterior pairs being on the plate and the third or posterior pair just off or barely touching it.

The **Straw Itch Mite** *(Pyemotes ventricosus)* — family Pyemotidae — is primarily predaceous on the larvae of grain insects such as Angoumois grain moth *(Sitotroga cerealella)* and the wheat jointworm *(Harmolita tritici),* but it will attack man readily. The body of the tiny mite is divided into cephalothorax and abdomen; the male is shorter and broader than the female. The gravid female becomes enormously distended, looking like a tiny pearl to the naked eye. Males and females are born alive and sexually mature. Farm workers and persons working with animals bedded down in infested straw are the ones most likely to be attacked. The resulting wheals, varying in size and form, are somewhat like those of chickenpox or scabies. Raised, blanched areas surrounded by rosy-red areoles result from rubbing. The lesions appear from ten to sixteen days after exposure; itching usually subsides in two or three days but may last for several weeks.

The **Itch Mite** *(Sarcoptes scabiei)* — family Sarcoptidae — is the cause of scabies or "seven-year itch" of man. The mite is round, with short legs, anal opening at tip of body (or slightly on the underside), pointed scales and blunt stout spines on the upper surface. Fertilized females cause most of the itching, by tunneling under the skin as they lay their eggs; males and the larvae and nymphs burrow less, being usually found on the surface or in hair follicles. Scabies is extremely infectious, the mites being readily transferred from one person to another where people are closely associated in work situations or otherwise. Infected persons become sensitized in a short time and quickly recognize a reinfection. [Scabies was reported (1971) to be on the rise in the Pacific coast states, in Mexico, Argentina, Brazil, Canada and eastern Europe; it was said to be "reaching epidemic proportions in some western European countries." There were similar reports of the United States as a whole in

1974.] Itch mites found on domestic animals cause mange or scab but are not readily transferred from one animal to another.

The **Follicle** or **Pore Mite** *(Demodex folliculorum)* — family Demodicidae — is found deep in the hair follicles and sebaceous skin glands of man, usually around the nose and eyelids; the scalp or other parts of the body may also be affected. People often have the mites without being aware of them. The infected area will show small, intensely red papules with white pinpoint tops, which contain the mites. They may be squeezed out and identified by placing on a microscope slide in a glycerin or chloral hydrate solution. All species of *Demodex* look much alike, are tiny, cigar-shaped, with abdomen ringed and legs very short. The entire life cycle is spent on the host. Closely related species attack domestic animals: the **Dog Follicle Mite** *(D. canis)* causes a mange seriously affecting dogs, the **Cat Follicle Mite** *(D. cati)* causes a mange on cats similar to that of dogs; the **Horse Follicle Mite** *(D. equi)* is a parasite of horses.

Ticks

Some ticks transmit disease organisms either as mechanical or as biological vectors — transmitting the organisms mechanically, by direct or fecal contact with the body orifices or skin abrasions, or through the mouthparts when the tick bites. Wood ticks — genus *Dermacentor* — are among the most important vectors of the organisms causing disease in wild and domestic animals and in man. With *Amblyomma* ticks, they transmit rickettsiae causing Rocky Mountain spotted fever, also the rickettsiae causing Q (or Query) fever; outbreaks of the latter have occurred among stockyard workers in the United States and Australia. *Dermacentor, Haemaphysalis* and *Ixodes* ticks transmit bacteria causing tularemia or rabbit fever among rabbits and rodents and sometimes to man. The bacteria of tularemia are passed from one generation of *Dermacentor* and *Haemaphysalis* ticks to the next via the egg — a phenomenon known as transovarial transmission — which serves to keep the microorganisms alive in an area. Soft ticks in the genus *Ornithodorus* transmit spirochetes causing relapsing fever from rodents to man. Tick paralysis (see p. 138), which occurs in man, dogs and cattle, is caused most commonly by the **Rocky Mountain Wood Tick** and the **American Dog Tick.**

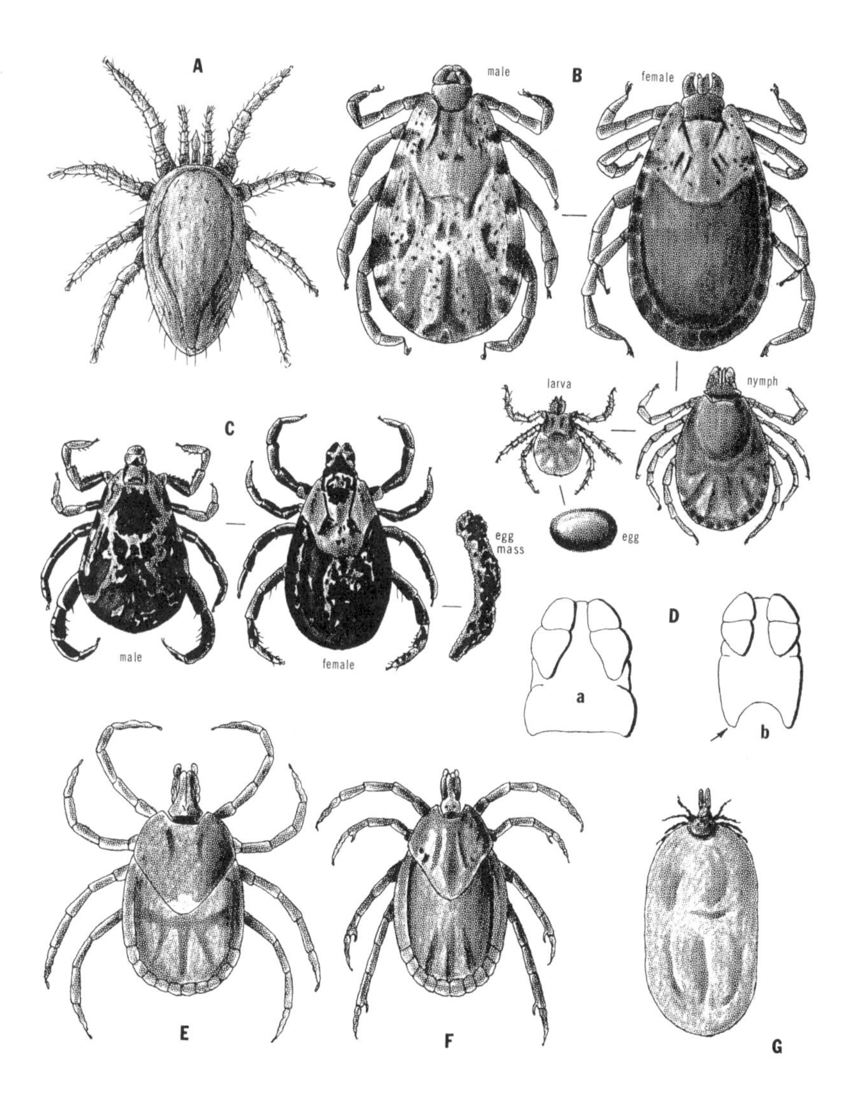

Fig. 54. Mite and ticks: A — northern fowl mite *(Ornithonyssus sylvarium)*; B —male and female of Rocky Mountain wood tick *(Dermacentor andersoni)*; C — American dog tick *(Dermacentor variabilis)*; D — capitulum of Rocky Mountain wood tick *(Dermacentor andersoni)* (a), and the Pacific Coast tick *(Dermacentor occidentalis)* (b); E — lone star tick *(Amblyomma americanum)*; F — Gulf Coast tick *(Amblyomma maculatum)*; G — black-legged tick *(Ixodes scapularis)*, engorged female.

There are two kinds of ticks, hard ticks (family Ixodidae) and soft ticks (family Argasidae). Hard ticks have a shield called the scutum, immediately behind the capitulum which projects from the anterior end of the body; the hard shield is small in the females and covers most of the body in the males.

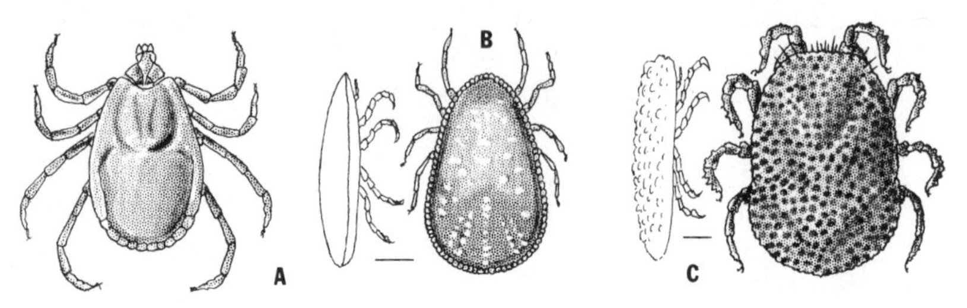

Fig. 55. Hard and soft ticks: A — brown dog tick *(Rhipicephalus sanguineus)*, a hard tick; B — fowl tick *(Argas persicus)*, a soft tick; C — a soft tick *(Ornithodorus moutaba)*.

Soft ticks do not have a shield and the capitulum projects from the underside of the body; both sexes of soft ticks resemble one another closely. Both the males and females of hard ticks are blood feeders but only the females become engorged; only one blood meal is normally taken by each of the feeding stages: larva (also called seed tick), nymph and adult. Ticks in the genus *Dermacentor* have a short capitulum and are easily removed from the flesh. Those in the genera *Amblyomma* and *Ixodes* have a long capitulum and are not so easily removed. Soft ticks, which are more common on birds than on man, are intermittant feeders; some species feed only at night and hide during the day, like bed bugs. A female hard tick will commonly lay several thousand eggs in a mass, extruding one at a time and covering it with a gelatinous substance. Soft ticks lay fewer eggs, in between periods of feeding.

The **Rocky Mountain Wood Tick** *(Dermacentor andersoni)* is found in most of the western states and in southwestern Canada (see distribution map). It requires two or three years to complete its life cycle. The larvae and nymphs feed on rodents mostly, adults get their blood meal from the larger mammals including man (94). The adult female is reddish brown, with a white shield, the male is mottled gray; the female expands from about 4.5 mm in length to about 12 mm when engorged, the male retains its original size. The engorged larva is about 1.5 mm long. Besides the organisms causing Rocky Mountain spotted fever, tularemia, Q fever and bovine anaplasmosis, this tick transmits a virus causing Colorado tick fever. The **American Dog Tick** *(D. variabilis)* occurs throughout the United States excepting the Rocky Mountain region (see distribution map), is most abundant in coastal areas and the Mississippi Valley. It requires from three months to two years to complete its life cycle, depending on the availability of hosts and the temperature. It is hardly distinguishable from the **Rocky Mountain Wood Tick,** but is somewhat darker in color; the adult female is slate-gray and about 12 mm long when engorged. The **American Dog Tick** is a vector of organisms causing the eastern form of Rocky Mountain spotted fever, tularemia and bovine anaplasmosis. [The **Winter Tick** *(Dermacentor albipictus),* which appears late in autumn, "is one of the most widely distributed ticks in North America," but it rarely bites man. The preferred hosts in nature are moose and deer; among domestic animals horses are often severely infested.]

The **Pacific Coast Tick** *(Dermacentor occidentalis)* takes the place of *D. andersoni* in the Pacific coastal region, may be distinguished by the tooth-like projections on the posterior margin of the capitulum. It is a serious pest of cattle, but is not very troublesome to man. The **Brown Dog Tick** *(Rhipicephalus*

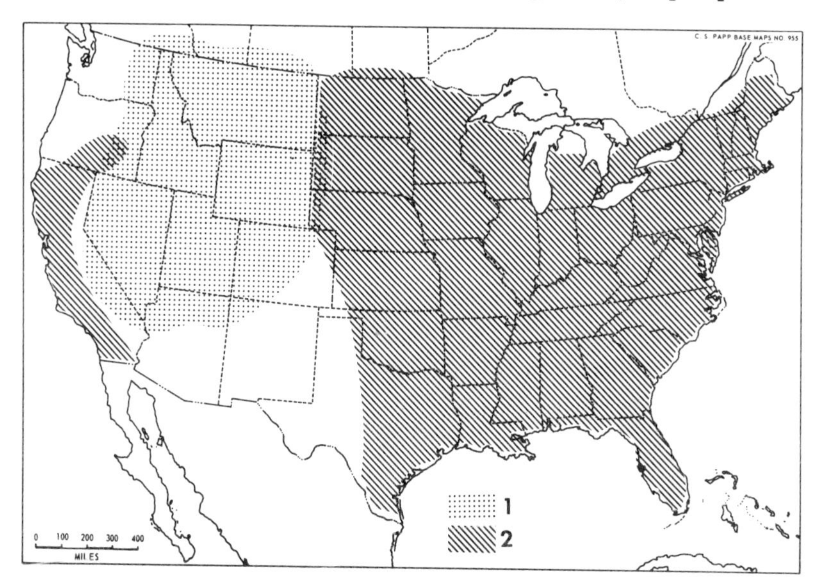

Fig. 56. Distribution of the (1) Rocky Mountain wood tick and (2) the American dog tick.

sanguineus), "one of the most widely distributed ticks in the world," is the tick most likely to be found in homes (where dogs are present); it will not survive the winters out of doors except in the southern states. The ticks are often attached to the ears or between the toes of dogs, but in severe infestations they are found on all parts of the body. Eggs and engorged ticks (the females swell to about 10 mm in length) may be found in cracks, on walls behind curtains, under furniture and rugs. They may bite man, but are not known to transmit disease organisms in this country. [According to Gregson it is "a potentially dangerous species." It transmits organisms causing boutonneuse fever among dogs and man in the Old World, and those of canine piroplasmosis, which occurs in many countries and has been found in the southeastern states. It is considered an important vector of the organisms causing Rocky Mountain spotted fever in Mexico.]

As noted above, ticks in the genus *Ixodes* have a long capitulum, consequently their bites are very painful and the mouthparts are not easily removed. They may be distinguished from other hard ticks by the anal groove, which extends from one side of the body to the other and curves around the front of the anus in a semicircle. The **Deer Tick** *(Ixodes pacificus)* is found only in the coastal region of British Columbia and western United States as far south as San Diego. The unengorged female may be recognized by the reddish brown body and the black legs, capitulum and scutum; it is 2.64 x 1.4 mm, and extends to 9 mm long when engorged. The male is black and about half the size of the unfed female. "The adults, which mate before feeding, attach to deer, dogs, cats, sheep and man, sometimes causing painful, slow-healing sores . . . The males attach briefly but repeatedly and leave irritating sores at the sites of their bites" (39). *I. pacificus* is very similar to the European **Castor Bean Tick** *(I. ricinus)* which it used to be called; it has also been erroneously called *I. californicus.*

The **Rabbit Tick** *(Haemaphysalis leporispalustris)* is widely distributed in the United States and Canada and is the tick most commonly found on rabbits. The unengorged female is about 2.5 x 1.5 mm, and extends to as much as 11.3 x 7.5 mm when engorged. Both male and female are dark brown or black, without light markings; the male is about half the size of the female, the latter slate colored when engorged. This tick is the principal vector of bacteria causing tularemia among rabbits and other wildlife. Most cases of tularemia in humans have been traced to contact with rabbits directly, or in some instances indirectly through cats that have become infected by rabbits; the transfer to humans is not ordinarily through bites of the tick. The **Bird Tick** *(H. chordeilis)* is widely distributed, similar in size and appearance to the **Rabbit Tick.** It occurs on rabbits, cattle, small birds and man, but is important mainly as a parasite of turkeys.

The **Fowl Tick** *(Argas persicus)* — a soft tick, known as the "blue bug" in the South — feeds on fowl at night and will bite man, causing severe pain and even shock. The male and female become engorged in an hour and require 30 to 40 days to complete their life cycle under favorable conditions. The upper surface of the adult has discs arranged in radiating lines, the upper and lower surfaces joined as it were by a seam (or impressed line) around the outer margin; it is reddish to purplish brown in color, from 8 to 10 mm long and half as wide. The larvae are reddish, with three pairs of legs until after the first molt when a fourth pair appears. The **Ear Tick** *(Otobius megnini)* is found throughout the western states and has become widely disseminated through cattle shipments; it has not however, established itself in any part of the United States except the arid and semiarid sections of the Southwest and Pacific coastal region. It feeds in the ears of horses, sheep, cattle and dogs. "Under favorable conditions they often attach themselves to other animals, such as goats, hogs and cats and occasionally even man." The adult is rounded at both ends and constricted near the middle, reddish brown to black, with short stiff spines, the female 7 to 8 mm long when engorged; all stages are without eyes. The engorged larva is grub-like, inactive, yellowish white or pink in color. Only the larval and nymphal stages are parasitic.

Soft ticks in the genus *Ornithodoros* are rounded and covered with "warts." They are usually found in rodent burrows and can transmit spirochetes causing relapsing fever if coming in contact with man; they can survive long periods without food and are apt to be encountered by hunters and campers when the ticks are most eager for a blood meal. The **Pahuello** *(Ornithodoros coriaceus)* occurs along the Pacific coast of California and Mexico, "has been taken from cattle and cattlemen," and has been collected on Mount Hamilton (Santa Clara County) "in deer beds among the low scrub oaks." The adult is reddish gray, with numerous deep pits, each bearing a hair; it is from 6.4 x 3.4 mm (male) to 13.8 x 8.2 (female). The larva is oval, convex dorsally, with sparse hairs and entire capitulum visible from above, about 3.7 x 2.75 mm when fed. Larvae and adults have two pairs of eyes. The **Relapsing Fever Tick** *(O. turicata)* occurs from California and Utah to Kansas, Oklahoma and Texas, in Florida and south to South America; it is light brown, with granular surface and no eyes, from 5.5 x 3.3 mm (male) to 9.6 x 6.8 mm (female). It attacks various hosts: reptiles, birds, rodents, rabbits, pigs, horses, cattle and man. The bite is said to be painless, "but followed in a few hours by intense local irritation and swelling. Subcutaneous nodules often form which, accompanied by occasional itching, may persist for months" (21).

Spotted Fever and Tick Paralysis

Like many other diseases, spotted fever is accompanied by a high body temperature, aching muscles and severe headache, but it is distinctive in producing a spotted rash (somewhat like measles) on the palms of the hands and soles of the feet. "It can almost always be cured with antibiotics (chlorophenicol or the tetracyclines) if diagnosed early enough;" misdiagnosis and treatment with sulfas and penicillin "seem to make it worse" (91). Cases of the disease occurring in Cape Cod and islands off the coast of Massachusetts were reported in "Time Magazine," January 24, 1969. One case outside the area was traced to a dog trained in Cape Cod and used elsewhere for duck hunting. Persons frequently exposed to wood ticks can get good protection against spotted fever by means of vaccination.

Tick paralysis is a strange malady, common only in the Pacific Northwest. It most frequently occurs in small children, with a high fatality rate. Paralysis symptoms — numbness of feet and legs, difficulty in walking and swallowing — appear five or six days after the female tick begins feeding; they are believed to result from the injection of neurotoxins secreted in the salivary glands of the feeding female tick (39). Complete recovery follows early removal of the ticks and the application of an antiseptic. Elsewhere, tick paralysis in animals and man is said to occur only in rare instances: when the ticks attack over the spinal cord or at the base of the skull. The case of a three-year-old girl turned up in Youngstown, Ohio not long ago (64); others may occur without being recognized. If spotted fever or tick paralysis is suspected, a doctor should be consulted as soon as possible.

First Aid and Prevention

Given some knowledge of ticks and mites, and with the exercise of a few precautionary measures, there need be no fear of going into areas infested with them. High-top shoes, leggings or socks pulled over the bottom of the pant-legs will keep ticks from crawling up the legs inside the trousers. Keeping the clothing buttoned and shirt-tail tucked into the trousers is "a must" in areas of exposure. The **Rocky Mountain Wood Tick** is abundant and active during the summer; the **American Dog Tick** is abundant in the northern part of the United States in the spring and early summer. In their southern range, the ticks are not so sharply influenced by seasonal changes.

Wood ticks may fall from trees or bushes onto people walking by but more often they are picked up in brushing against low vegetation. When hiking through wooded areas and fields infested with ticks, it is advisable to examine the clothing occasionally (with one person checking the other), so as to remove any that may be present before they become attached. Sitting on the ground or a rotten log in woods or fields harboring ticks or mites is a sure way to become infested with them. (Large rocks or boulders are a safer place to sit in woods or fields.) Clearing brush along paths and keeping weeds and grass cut in recreation areas will help prevent infestations of ticks and mites.

After exposure to a tick-infested area and before bathing, one should look for ticks on the body, especially the head, base of skull and pubic region. When attached to the skin, they should be *lifted* off with a slow, gentle pull to remove the mouthparts (if preserved these will aid in identification). Ticks with long mouthparts *(Ixodes* and *Amblyomma)* may require a twisting motion with the pull, or the application of an irritant, such as the tip of a hot needle (some persons use the end of a lighted cigarette), or a drop of turpentine, gasoline, benzine or carbon tetrachloride. Vaseline or fingernail polish rubbed over the tick will sometimes help to remove it. If none of these methods is effective, and the mouthparts remain, they should be removed by an incision, and an antiseptic applied. Spotted fever will not develop unless *Dermacentor* ticks are attached for at least two hours or more (both males and females are infective). As already stated, these ticks are usually easy to remove because of their short mouthparts; they attach themselves to the skin by means of a rapidly-hardening cement, which pulls off with the tick and has the appearance of a skin papule. If fever developes after a tick bite, one should consult a physician, telling him of the bite; having a specimen of the biter will help in diagnosis and treatment.

In general, the same protective measures outlined for ticks will apply to mites. Chiggers are probably the most troublesome of the mites to man. The first warning of an attack may not be until the welts begin to form and itching starts (see p. 128). When the infestation is discovered, one should take a bath as soon as possible, applying a thick lather of soap and rinsing off several times. The sooner the mites are discovered, the better will be the effect of bathing. The bath will kill most of the chiggers; applying a dab of alcohol or antiseptic to each welt will destroy any that are still alive. Proprietary ointments containing benzocaine for relief of chigger and other bites are available. Prolonged use of compounds containing benzocaine, which has an anesthetizing effect, should be avoided.

Repellents

Repellents — generally the same formulations as used against fleas, flies and mosquitoes — provide good protection against ticks and mites. They are longer lasting and more effective if applied to the clothing as well as the skin. They may be applied lightly to exposed parts of the body — neck, arms and legs — with the fingers or other means; application to the entire body is impractical and may be injurious. As stated before, diethyl toulamide (DT) is one of the most effective repellents against biting flies, fleas, ticks and mites. Ethyl hexanediol (EH) alone or in combination with dimethyl phthalate (DMP) and Indalone, is said to be an effective repellent and "in general nonobjectionable to use, but it should be kept away from the eyes and mouth and must not be permitted to come in contact with plastics, for which it is a solvent" (40). Rayon and some other manmade fabrics are said to be harmed by the repellents but not nylon, cotton or wool (95).

Most brands of insect repellents contain DT or EH as the active ingredients, in varying concentrations. DT is 50 percent more effective than EH, hence the latter required greater potency for the same effect. Concentrations of DT in various brands of insect repellents varies from a few percent (having little value) to 50 percent; EH concentrations range up to 75 percent to compensate for it lesser effectiveness. Compounds with greater amounts of active ingredients are naturally more effective and longer lasting. Various forms of repellents are available: liquid, spray, cream, foam, stick and towelette. Liquids, in general, contain more active ingredients, hence require lesser amounts and less frequency of application to be effective. Sprays have the advantage of being easier to apply to the clothing, but will require more frequent application. It should be understood that these applications do not keep the insects from flying around you or touching the skin. It is contact with the repellent that prevents them from biting — the *touch,* and not the smell of it (66).

Repellents are most effective if applied to clothing around the upper edge of the socks, on trouser cuffs and waistband, dress hem and waistband, shirt cuffs and neckband. Impregnating (soaking) the clothing with DMP or benzyl benzoate will give longer-lasting protection; the latter is effective longer than the other since it does not wash out as readily.

Summary of Insects and Other Arthropods That Attack Man

Insect or Other Arthropod	Dominant Colors	Average Size	When Active				Type of Injury			Site of Attack			Local Reaction					Residual Effect of Injury
													Effect		Duration			
			Day	Night	Dawn	Dusk	Sting	Bite	Nettles	Single	Scatter	Group	Immediate	Delayed	Hours	Days	Weeks	
Honey bee	Black, yellow or gray bands	12-20 mm.	X				X			X			X	X	X	X		None
Bumble bee	Black and yellow	12-25 mm.	X				X			X			X	X	X	X		None
Yellow jackets (paper-nest wasps)	Black, yellow	12-20 mm.	X				X			X			X		X	X		None
Bald-faced hornet (paper-nest wasps)	Black and white or pale yellow	12-20 mm.	X				X			X			X		X	X		None
Polistes wasps (paper-wasps)	Black and yellow	12-25 mm.	X				X			X			X		X			None
Mud daubers (thread-waisted wasps)	Black and yellow, or blue	12-25 mm.	X				X			X			X			X		None
Tarantula hawk (spider wasp)	Metallic blue, fiery red wings	32 mm.	X				X			X			X			X		None

Velvet-ants (mutlid wasps)	Black, dense white or red hairs	12-25 mm.	X		X			X			X	X			None
Legionary or army ant	Reddish brown	3-12 mm.	X	X	X	X		X			X		X		None
Harvester ants	Reddish brown	6-12 mm.	X		X	X		X			X		X		Severe swelling
Fire ants	Yellowish to reddish	3-6 mm.	X		X	X				X	X		X		Pustule, crust, scar
Carpenter ants	Black or brown	8-12 mm.	X			X		X			X	X			None
Silverspotted tiger moth caterpillar	Black and brownish dense hairs	38 mm. full grown	X				X		X		X	X	X		Reddish papules
Nevada buck moth caterpillar	Greenish, with brown spots	40 mm. full grown	X				X		X		X	X	X		Rash
Western tussock moth caterpillar	Gray; red, yellow, blue spots; white, black tufts	13-25 mm. full grown	X				X		X		X	X			None
Range caterpillar	Yellowish, gray or black	60 mm. full grown	X				X		X		X		X	X	Papules; asthmatic, catarrhal symptoms
Blister (meloid) beetles	Metallic green and purplish, black or brown	8-27 mm.	X								X		X		Blisters

Bloodsucking conenose bug	Brownish to black	20 mm.		X			X		X	X		X		X	None
Western corsair (bug)	Amber	5 mm.		X			X	X			X		X		None
Wheel bug	Brown	32 mm.	X				X	X			X		X		None
Giant water bugs	Brown	25-40 mm.					X	X			X		X		None
Backswimmers (bugs)	Black and white	12-20 mm.					X	X			X		X		None
Water-scorpions (bugs)	Brown	12-25 mm.					X	X			X		X		None
Bed bug	Reddish brown	6 mm.		X			X			X	X		X		None
Fleas	Light to dark brown	2-3 mm.					X		X		X		X	X	None
Body louse	Grayish white	2-4 mm.	X				X		X		X		X		None
Head louse	Grayish	Minute (2 mm. or less)					X		X		X		X	X	None
Common mosquito	Brown or black yellowish bands; proboscis, legs white-ringed	4 mm.		X	X	X	X	X			X		X		None

Biting midges	Brownish gray, brown, or black	2-3 mm.		X	X	X		X		X		X		X	Sometimes nodule
Black flies	Black to grayish brown or yellowish	4-6 mm.	X	X		X		X			X			X	Nodule, scar
Horse flies	Black or slate-gray or yellowish brown	20-30 mm.	X			X	X			X		X			None
Deer flies	Black and yellow	12 mm.	X			X	X	X		X		X			None
Stable fly	Gray and yellowish	6-10 mm.	X			X	X			X		X			None
Eye gnats (Hippelates flies)	Black to gray	Minute (2 mm. or less)	X					X		X			X		Conjunc-tivitis
Pigeon fly (louse fly)	Brown; wings and legs yellowish	6.5 mm.	X			X	X			X		X			None
Chiggers (mite)	Orange-yellow or red	Microscopic	X			X			X		X		X		Reddish welt
Straw itch mite	Pearl-like (gravid female)	Minute (2 mm. or less)	X			X			X		X		X	X	Reddish wheal

Adapted in part from Frazier (31).

Summary of Mites and Ticks Known to Carry Disease Organisms Affecting Man in the United States

Common name:	Scientific name:	Distribution:	Disease:	Disease organism:
Mouse Mite	*Allodermanysus sanguineus*	Throughout the U.S.	rickettsial-pox	rickettsia: *Rickettsia akari*
Rocky Mountain Wood Tick	*Dermacentor andersoni*	Most western states, southwestern Canada	Rocky Mountain spotted fever	rickettsia: *Rickettsia rickettsii*
			Q fever	rickettsia: *Coxiella burnetti*
			tularemia (rabbit fever)	bacterium: *Pasteurella tularensis*
			Colorado tick fever	virus
American Dog Tick	*Dermacentor variabilis*	East of the Rocky Mtn. and in Oregon and California	Rocky Mountain spotted fever	see above
			tularemia	see above
Pacific Coast Tick	*Dermacentor occidentalis*	Pacific Coast	Rocky Mountain spotted fever and tularemia suspected	see above
Lone Star Tick	*Amblyomma americanum*	Texas eastward, north to Iowa	Rocky Mountain spotted fever	see above
			tularemia	see above
			Q fever	see above
Soft Ticks	*Ornithodoros* (several species)	Florida to California, northward in the western states	ralapsing fever	spirochetes: *Borrelia* spp.

From a leaflet entitled *Precautions Against Arthropod Venom and Disease*, provided as a public service by the Automobile Club of Southern California and the San Diego Natural History Museum.

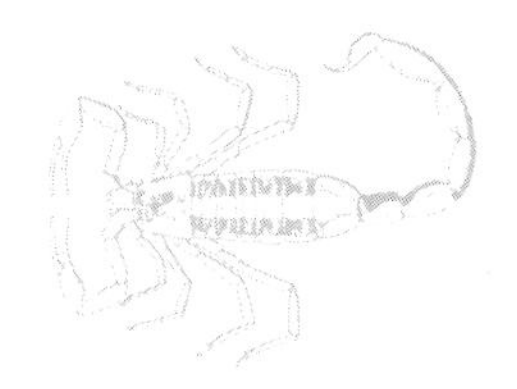

IX. SPIDERS, SCORPIONS AND CENTIPEDES

Besides the mites and ticks (order Acarina), the class Arachnida includes the following: spiders (order Araneida), daddylonglegs and harvestmen (order Phalangidae), scorpions (order Scorpionida), whipscorpions (order Pedipalpida or Uropygida), pseudoscorpions (order Chelonethida or Pseudoscorpionida), solpugids (order Solpugida) and some small groups. (See pages 125 and 150 for distinguishing features of the arachnids.) Daddylonglegs differ from spiders in having the cephalothorax and abdomen broadly joined and in lacking silk glands; they are found in fields everywhere and prey on small insects, but do not bite humans.

Whipscorpions, False Scorpions and Windscorpions

Scorpions have two lobster-like pincers (pedipalps) and long, segmented bodies ending with a stinger. Whipscorpions are similar, with these differences: the pedipalps are developed into stout pincers, the inner edges armed with teeth and spines, for grasping and crushing their prey; the front legs are long and slender, having a sensory rather than walking function; a stinger is lacking. Some whipscorpions have a whip-like tail, others are without a tail. They are sometimes called "vinegaroons" because some tailed species emit a strong odor of vinegar when using their defensive spray, which has been found to contain acetic acid and caprylic acid and to originate at the base of the slender tail or telson. Large species such as the **Giant Whipscorpion** *(Mastigoproctus giganteus)* — family Thelyphonidae — are able to inflict a painful wound with their powerful pincers, but they are not venomous. The **Giant Whip-scorpion,** which occurs throughout the southern states, is reddish brown and often attains a length of 75 mm; Florida specimens of the female, which tend to be smaller than those found in other regions, are from 38 to 50 mm long. Whipscorpions are predaceous on insects and other small animals, and are usually found under logs, boards, rocks, debris and other places, but occasionally get into houses. In the tailless whip-scorpions, the front legs are extremely thin and long. Pseudo-scorpions — known as false scorpions — are very small

Fig. 57. A Hadrurus scorpion, native to the American deserts (top); stinger of the same (bottom).

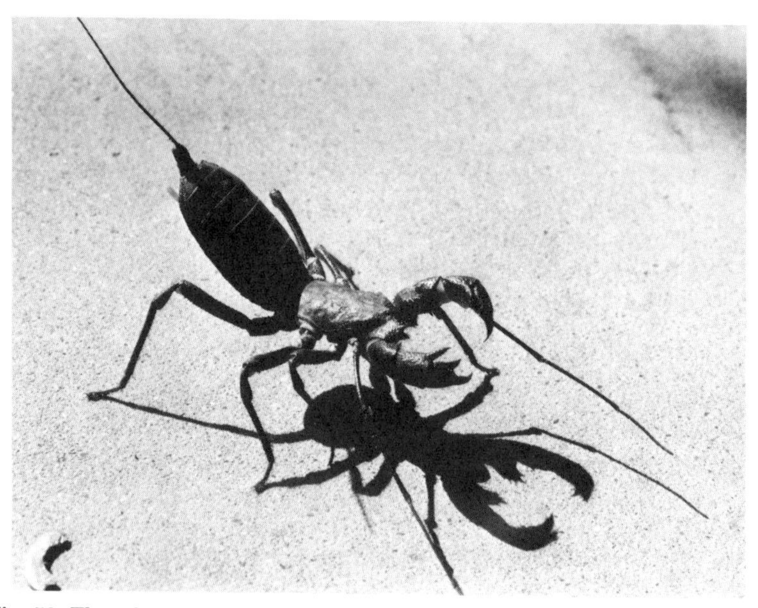

Fig. 58. The whip scorpion is harmless. Note the uniformly broad abdomen, with the long, whip-like extremity.

arachnids, being less than 6 mm long, and also resemble scorpions, except that they have no tail, nor stinger. They are harmless creatures, and abundant in wooded regions throughout the world; according to Savory, about 2,000 species constitute the order (77).

Solpugids — sometimes called sunspiders or windscorpions — appear to have three body segments instead of two, and lack pincers; they have four pairs of walking legs and a pair of leg-like adhesive pedipalps. Windscorpions are medium to large in size (the largest North American species is about 75 mm long) and have strong mouthparts capable of inflicting a severe bite, but they are not venemous. Their habitat is the hot deserts where they hide in burrows or crevices during the day and come out to feed at night; being attracted to lights, they may be encountered in this way. *Eremorhax magnus* (family Eremobatidae) is found in California, Nevada, Arizona, New Mexico, Texas and Mexico, averages about 36 mm in length. *E. striatus* is very similar, occurs in Nevada and California, may be distinguished by the darkened areas on the femora and tibiae of the legs and pedipalps; it is about 38 mm long.

Spiders

Perhaps no other familiar creature has been so generally abhorred or has elicited such exaggerated fear as the spider. Shakespeare expressed this loathing and fear of spiders in *The Winter's Tale,* thus:

> There may be in a cup
> A spider steep'd, and one may drink, depart,
> And yet partake no venom, for his knowledge
> Is not infected; but if one present
> The abhorr'd ingredient to his eye, make known
> How he hath drunk, he cracks his gorge, his sides
> With violent hefts.

Spiders are retiring by nature and will not normally attack unless cornered or aggravated; this goes for the black widow spiders as well as the harmless ones. The only exception is thought to be a funnelweb spider found in Australia. The touching of a spider's web brings a quick movement in that direction since this is the way the web spinners capture and immobilize their prey; their vision is poor and they respond more to movement. Exceptions are the hunters, wolf spiders (family Lycosidae), and jumping spiders (Salticidae) which "have the keenest vision of all spiders" (48). Spiders are predaceous on insects and while not choosy as to their prey, they are considered beneficial on the whole and important in maintaining a balance among the species. Excepting two small groups, they all have venomous glands but not many of them have sufficiently powerful "jaws" (chelicerae) to plung the fangs into the human skin.

Spiders have a thread-waisted joint or pedicel joining the cephalothorax and abdomen, and four pairs of walking legs, which have seven segments and are joined to the cephalothorax (never to the abdomen). There is no larval stage as in ticks and mites (see p. 125); the young or spiderlings look like the adults except in size, and have the same number of legs. Like the scorpions and their close relatives, spiders are equipped with another pair of appendages called pedipalps; they resemble the walking legs but have six segments, are in front of the walking legs, attached to the underside of the head behind the mouth, and are sensory and prehensile in function. The terminal segment or tarsus of the pedipalps is bulbous in all male spiders

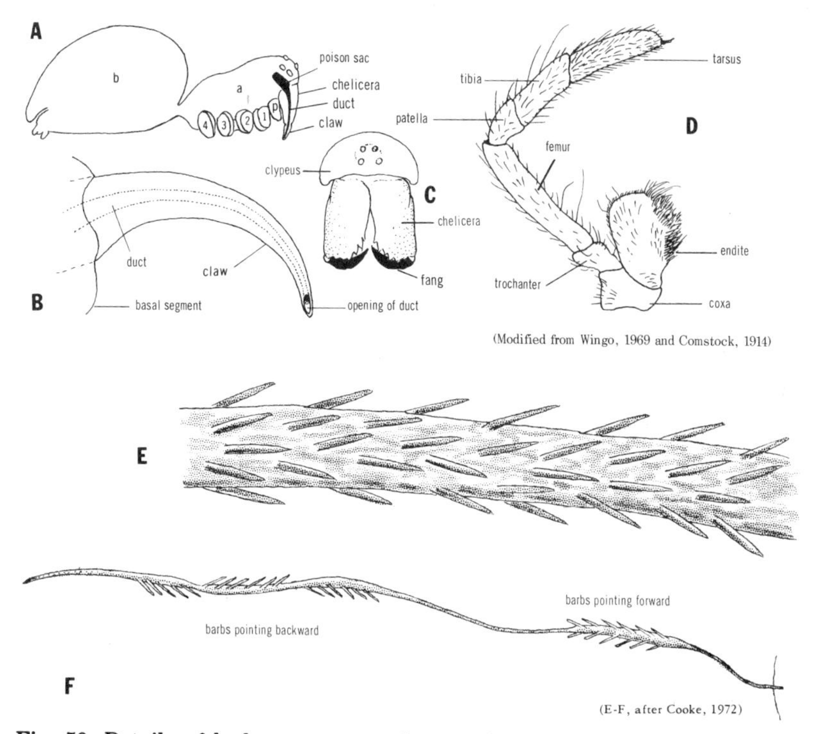

Fig. 59. Details of body structure and appendages of a spider: A — cross-sectional outline of body showing cephalothorax (a), with poison sac, chelicera, and poison duct ending in claw or fang and posterior of palp (p), legs (1, 2, 3, 4) and abodmen (b); B — claw or fang of brown recluse spider (note *lateral* opening of duct which prevents clogging by tissues of prey); C — face of spider showing chelicera and fangs; D — pedipalp (of female) showing six segments: coxa, trochanter, femur, patella, tibia and tarsus (the *leg* has an additional segment, the pretarsus, between the tibia and tarsus; the endite or lobed structure is not a segment). Stinging hairs of tarantula; E — type found on Mexican tarantulas and which produces severe skin irritation; F — type found on tarantulas occuring in the U.S. and which causes a relatively mild skin reaction.

and modified as a copulatory organ; it is simple and thread-like in the female. The swollen palpal tarsus of the mature male provides an easy means of distinguishing the sexes; the male may be recognized in most cases one molt prior to maturity, in some species two molts. The bulb of the male palps contains the embolus, a more or less complicated organ which is withdrawn into a spirally wound coil. The palpal organ is charged with semen (from the testes located in the abdomen and opening on the underside) before courtship and mating takes place; the

Fig. 60. Tarantulas. Top: female of a species *(Aphonopelma)* found in our Southwest, guarding her young. Bottom: a species *(Brachypelma)* found in Mexico.

emboli are inserted in the female genitalia — located on the underside of the abdomen — during mating.

The eyes of spiders are simple (ocelli), usually eight in number and sometimes arranged in curving rows. In front of the mouth and above it is a pair of chelicerae or "jaws." The chelicera is a two-segmented appendage consisting of a basal segment and the claw. The claw ends in a sharp point and has an opening leading through the duct to the venom gland, which lies in the basal segment of the chelicera or extends beyond it, according to the species. When the claws are pressed *laterally* into the prey, the venom runs through the ducts into the wound.

Tarantulas

Tarantulas (family Theraphosidae or Aviculariidae) are generally considered harmless; their bite is said to "feel like a pinprick" (102), followed by mild stinging pain and soreness. [Dr. Arthur C. Smith takes exception to this, having been painfully bitten by a small tarantula; prolonged soreness and partial paralysis of the affected arm followed. Venom from tarantulas *(Aphenopelma* sp.) was found by Stahnke to be five percent protein, compared to 62 percent as found in scorpions *(Centruroides sculpturatus);* lethality of the two venoms on rats was in the same proportion.] Unlike other spiders, their jaws move vertically; to bite they must raise the front of the body up and plunge their fangs downward. Tarantulas found in the United States have short, kinked, barbed urticating hairs which may cause itching in some persons when coming in contact with their skin. Some species found in Central and South America have longer barbed hairs that cause a much more severe reaction. Those of "a colorful Mexican species . . . sometimes sold in New York pet stores . . . produce a large and persistent rash on human skin" (20). The hairs are flicked off the tarantula's abdomen into the face of rodents or other animals attacking them, which usually discourages their adversaries.

"There are about 30 species in the family," according to Kaston, two-thirds of them being in the genus *Aphonopelma;* they occur mostly in the Southwest (48). It is difficult to tell one from the other; females are usually brownish, the males darker, often almost black, with rust-colored hairs covering the abdomen. *Aphonopelma eutylenum* is a dark brown to blackish species, from 40 to 43 mm long, occurs in southern California, Arizona and New Mexico. *A. baileyi* and *reversum* are similar and found in the same area. Tarantulas are mostly nocturnal, hiding during the day in their own burrows or those abandoned by rodents and under rocks. Those seen wandering in the open are usually the males and only during the mating season; this is during October and November in southern California. It takes ten years for a young tarantula to mature, if it is lucky enough to escape its predatory and parasitic enemies and other hazards; in this event they have the chance of adding another ten years or more to their life span.

The Harmful Spiders

Only two kinds of spiders are considered dangerous: the black widows *(Latrodectus)* and violin spiders *(Loxosceles);* only black widows are able to inject "sufficient venom to cause death" (79). The venom of a black widow spider is said to be 15 times as toxic as that of the prairie rattlesnake, but since the quantity injected with a single bite is minute, the resultant mortaility is less than 1 percent compared to 15 to 20 percent mortality resulting from rattlesnake bites. The venom of black widows is neurotoxic (deadens the nerves) while that of the violin spider is necrotic (kills tissues) in effect. Only females of black widows bite; both males and females of violin spiders can inflict a venomous bite. Occasionally black widow spiders are ingested with food but "they are not poisonous under these circumstances" (79).

Black Widow Spiders

Three species of black widow spiders are recognized by Kaston as occurring in the United States (and partially in Canada): *Latrodectus mactans,* which is southerly in distribution; *L. variolus,* which extends farther north and into Canada; and *L. hesperus,* which occurs west from about the middle of Texas, Oklahoma and Kansas, and north into Canada (27). The "brown widow" *(L. geometricus),* and the "red widow" *(L.*

bishopi) are limited to southern Florida. The female of the **Black Widow** [According to "Common Names of Insects" (Ent. Soc. Amer.), the name **Black Widow** applies specifically to *Latrodectus mactans.*] *(L. mactans)* is shiny black, smallest of the three, from 8 to 10 mm long, with a red hourglass mark on the underside of the abdomen and a row of small red spots down the middle of the back; the hourglass mark takes various shapes and is sometimes indistinct. The male is much smaller — about 5 mm long — with a white band encircling the front end of the abdomen and two diagonal white bands on the sides and a median row of broad, red spots on top of the abdomen; in some specimens the legs are ringed with a lighter color. The female of *L. variolus* is black, with large red spots down the middle of the back, three narrow white stripes on the sides and one encircling the front of the abdomen; the hourglass mark is divided. The male has four diagonal white stripes on the sides. The female of the **Western Black Widow** *(L. hesperus)* is slightly larger than the other two, usually all black, with the

Fig. 61. Black widow spider: Female (left) and male (right). (USDA photo).

hourglass ordinarily complete and narrowly constricted in the middle. The background coloring of the male is much lighter than in *mactans,* with the white markings on the sides similar except that the anterior stripe is hooked at the ends.

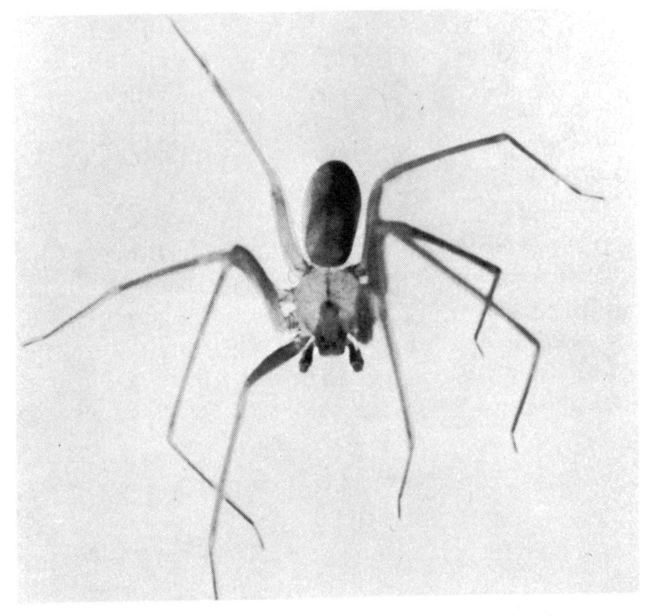

Fig. 62. Brown recluse spider *(Loxosceles reclusa),* also called violin spider.

Violin Spiders

The **Brown Recluse** or **Violin Spider** *(Loxosceles reclusa)* is slightly smaller than the female **Black Widow,** being about 9 mm long (the male slightly smaller than the female). It varies from light fawn to dark brown in color distinctly marked with a darker violin-shaped band extending from the eyes nearly to the base of the cephalothorax. The eyes are six in number. Like the black widow spiders, it is an outdoor spider but is often found around houses and out-buildings, and spins an irregular web (41). Geographically, it is concentrated in Arkansas, Kansas, Oklahoma and Missouri, but is also found in adjoining states. The bite of violin spiders *(Loxosceles)* is often insidious, the victim not being aware of having been bitten for several hours.

A large species *(Loxosceles laeta)* closely related to the **Brown Recluse** and indigenous to Brazil and Chile was found in southern California in 1969. It was referred to in the press as the **Violin Spider,** and said to be more deadly than the **Black Widow** (52). The infestation was quite large, and determined attempts to eradicate it apparently have not been successful. The venom of the two western species of *Loxosceles — unicolor* and *arizonica* — is less virulent than that of *reclusa* and *laeta. L. unicolor* occurs in Texas and Colorado, west to California; the cephalothorax is yellowish orange, with only a faint violin-shaped mark in some specimens, the abdomen is brownish gray, the female about 7 mm long. *L. arizonica* resembles *unicolor* closely, occurs in Arizona and New Mexico.

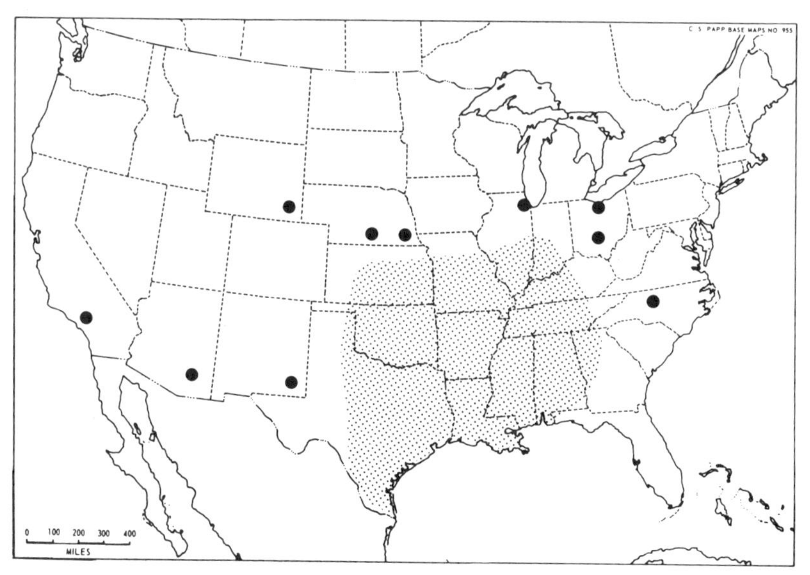

Fig. 63. Distribution of the brown recluse spider.

Spider Bites: First Aid and Prevention

The recommended first aid treatment for a spider bite is to apply "an antiseptic solution to the puncture made by the spider's fangs to prevent infection, and ice packs to localize the venom" (102). The victim should be kept as quiet as possible and taken to a physician or hospital without delay. A tourniquet is not recommended as in the case of scorpion stings or snake bites since the venom acts instantaneously. A specific antivenin is available for black widow bites, which are characterized by the presence of fang puncture marks and surrounding redness. The bite of a black widow spider is described as being like a pin prick, followed shortly by an intense burning sensation and excruciating pain; this may be followed by symptoms similar to those of appendicitis: severe cramps and rigid abdominal muscles, nausea, excessive perspiration, rise in body temperature, difficulty in breathing and talking.

No antivenin is presently available to counteract the venom of the violin spiders. The puncture site in this case becomes temporarily reddened, followed quickly by blistering; fang puncture marks are rarely evident. The reaction to the bite of a violin spider may be mild or it may be very severe. Typically, "the bitten part becomes painful and swollen, and blisters often form on the skin around the bite. The next day the skin at the bite site begins to turn purple; during the next week or so the skin turns black as the cells die. Several weeks later the blackened area falls away, leaving a circular pit in the skin. This pit slowly fills with scar tissue. In a few persons, the venom of the brown recluse spider has caused destruction (hemolysis) of many red blood cells, a very serious complication signaled by the appearance of bloody or dark-colored urine" (37). A few fatalities due to the bite have been documented.

Spiders are normally active at night and hide during the day in trash, closets, bedding and other secluded places. Closets and storage areas should be cleaned out occasionally and clothing and bedding should be shaken before using. Spiders are most active in the spring and early summer. Spider bites usually occur in and around living quarters and adjoining buildings, when people rub against webs or reach blindly into dark corners, and when they are putting on clothing. Most deaths in the United States have occurred among males, bitten on the genitals while using outdoor toilets; most of the victims have been migrant farm workers in California.

Scorpions

Scorpions are closely related to spiders; like them, they have four pairs of legs attached to the cephalothorax (with one less leg segment than in spiders), pedipalps (they have one less segment than in spiders, and have pincers) and chelicerae (which are small, three-segmented, and have pincer-like claws). The distinctive feature is the abdomen, consisting of a broad preabdomen of seven segments and a narrow postabdomen of five segments, not counting the bulbous terminal segment (or telson), which contains two venom glands and ends in a sharp stinger. Another feature peculiar to scorpions is the pair of comb-like structures called pectines lying on the underside of the cephalothorax and attached between the bases of the last pair of legs; they are thought to serve a sensory function.

Scorpions are nocturnal and rarely seen moving about during the day. Their food consists of insects for the most part but includes spiders, centipedes, earthworms and other small animals, but not sowbugs. They are not aggressive and do not stalk their prey, their vision being poor; they move about or lie in wait, relying on the sense of touch transmitted through the hairs on the body and pincers. Unlike spiders, which lay eggs in sacs, scorpions give birth to living young. They are born in semitransparent envelopes (which have been mistaken for egg sacs by some observers); the young break out themselves and rarely require the assistance of the mother. They are carried about on her back for several days, until they molt; it takes them about a year or more (depending on the species) to reach maturity. The female scorpion often eats her mate and frequently her young when hungry, a savagery toward her mate that belies her courtship manner. Scorpions have attracted much attention by their interesting mating behavior, which has amused and often puzzled persons who have witnessed it. The ritual consists of the male and female facing one another, with "arms" (pedipalps) outstretched and pincers locked, and performing a *grand pas de deux* consisting of movements forwards, backwards and sideways. This "mating dance" serves as a means of transferring the male sperm to the female without

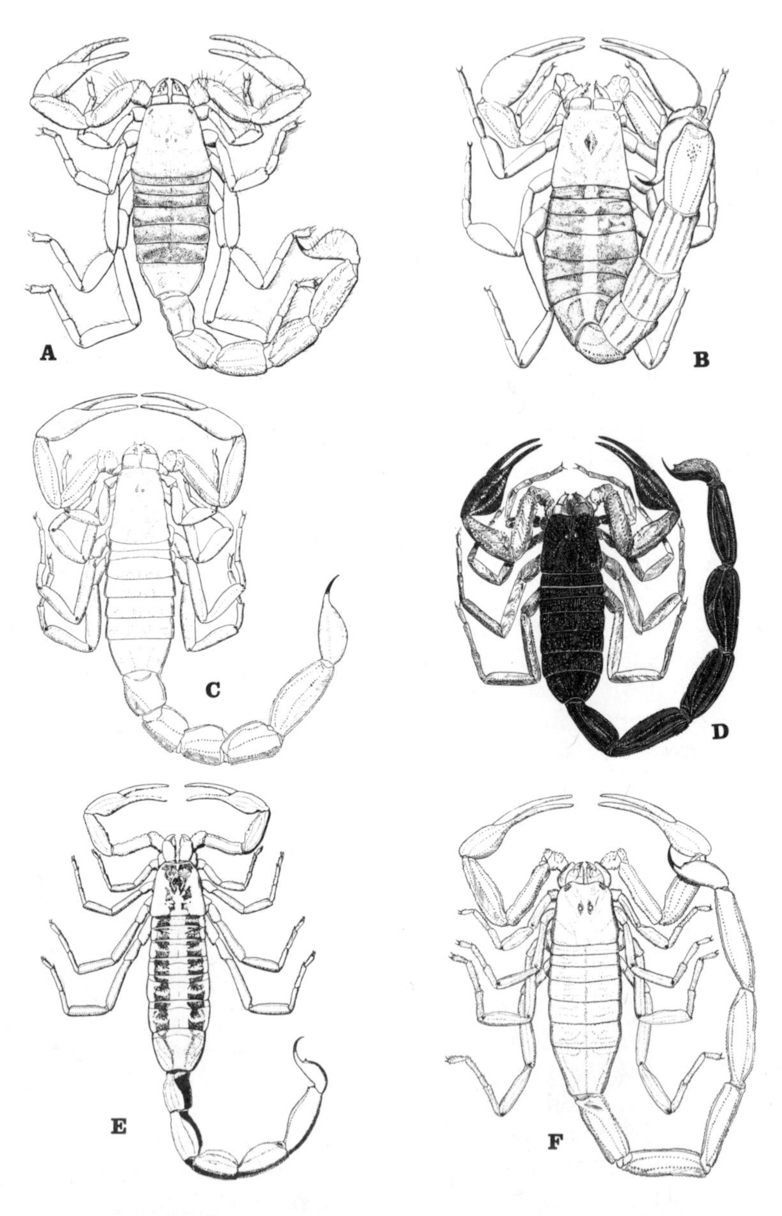

Fig. 64. Scoropions: A — olive hairy scorpion *(Hadrurus arizonensis);* B — stripe-tail devil scorpion *(Vejovis spinigerus);* C — slender devil scorpion *(Vejovis flavus);* D — black scorpion *(Centruroides marginatus);* E — stripe-back scorpion *(Centruroides vittatus);* F — deadly sculptured scorpion *(Centruroides sculpturatus).* (US. PHS.)

direct copulation; he glues the encapsulated sperm to the ground (after finding a suitable spot with these movements) and maneuvers her over it so it can be picked up in her genital opening.

Adult scorpions vary in length from something under 25 mm to about 200 mm, and in color from nearly black to straw; their bodies are usually smooth and some are hairy. Our lethal

Fig. 65. Scorpions in their natural habitats.

species — *Centruroides sculpturatus (gertschi)* — found in Arizona, is greenish yellow, lemon-yellow or straw color, and has a blunt thorn called the subaculear spine or tubercle at the base of the stinger (most species of *Centruroides* have a sharp spine). The nonlethal species are blackish, brown, or gray and do not have the subaculear spine, excepting some *Centruroides*. Scorpions are most common in Texas, Arizona and the southern part of California, Nevada and Utah, but occur in most states east of these, to the Atlantic seaboard. The **Deadly Sculptured** or **Deadly Striped Scorpion** *(Centruroides sculpturatus)* — family Buthidae — is uniformly yellowish or straw-colored, about 70 mm long; the female is thick-bodied, the male more slender and with longer tail. The form *gertschi,* previously considered a separate species, has two broad, irregular black stripes down the back. According to Stahnke (85), the sting of *C. sculpturatus* produces "no local swelling and the effects are principally systemic"; other species of *Centruroides* found in the United States "have a venom that may produce a very painful and pronounced local swelling with very mild, or no noticeable systemic effects." (For systemic reactions to *C. sculpturatus* stings, see p. 164).

The **Margarite Scorpion** *(Centruroides gracilis)* occurs from Florida to California, south to Brazil and Chile, and in West Africa; it is usually dark brown or black, without stripes or subaculear spine, about 80 mm long. The **Common Striped Scorpion** *(C. vittatus)* — the most common and widespread of our scorpions — occurs from New Mexico to Florida and South Carolina, north to Tennessee, Kentucky, Missouri and Kansas, south to Mexico; it is yellowish brown, with two broad, black irregular stripes down the middle of the abdomen, interocular triangle darker than surrounding area of cephalothorax, stinger with sharp suboculear spine, length of body about 65 mm. According to Baerg (4), the sting of this species "results in a sharp pain that usually lasts for 15 to 20 minutes. Persons stung in late March or early April, when scorpions have just emerged from their winter quarters, may experience a more severe pain lasting for several hours, and in addition some lameness in the tongue plus a general numb feeling."

The **Stripe-Tail Devil Scorpion** *(Vejovis spinigerus)* — family Vejovidae — occurs from California to Texas, southward to Mexico; it is greenish or yellowish, with broad, faint yellowish median stripe on upper side of preabdomen, and four longitudinal stripes on underside of postabdomen, from 50 to 80 mm long. Baerg, who allowed this species to sting him, described the sensation as "very much like that of a pin prick, and the resulting pain, which was very slight, lasted scarcely half an hour. There was no white area around the punctures and not the slightest swelling or inflammation." The **Mordant Scorpion** *(Uroctonus mordax)* occurs in California and Oregon, is also reported from Kentucky; it is dark brown, the cephalothorax and pedipalps darker than the preabdomen, the latter darker than the legs, has large pedipalps and stout pincers, and a somewhat reduced postabdomen. Like most other vejovids, its sting usually causes only transient pain, with occasional swelling and tenderness.

The giant hairy scorpions, genus *Hadrurus* (family Vejovidae), are among our largest scorpions, the bodies of mature specimens being from 100 to 127 mm long; they are heavy-bodied, and conspicuously hirsute. The **Black Giant Hairy Scorpion** *(Hadrurus spadix)* occurs in the arid regions of California, Oregon, Idaho, Nevada, Utah, Arizona and Colorado; the cephalothorax and preabdomen are dark olive to black dorsally, the postabdomen, stinger and legs yellow, mature specimens up to 107 mm long. It burrows in the loose soil and under rocks and other objects. The **Arizona Giant Hairy Scorpion** *(H. arizonensis)* is found in the Sonoran Desert of Arizona, northward to Death Valley in California, southern Nevada and Utah, and southward to Mexico; a very large species, it may attain a length of 127 mm; the cephalothorax and preabdomen are dark olive in color, excepting yellow markings in the interocular area. Despite their size, the stings of *Hadrurus* usually produce only localized pain and some swelling, with occasional discoloration around the wound.

Scorpion Stings: First Aid and Prevention

The reaction to scorpion stings is lighly variable and depends on the species of scorpion, the amount of venom injected and the sensitivity of the person involved. Most nonlethal species, as suggested in our discussion of individual species, produce only local and transient symptoms: pain, sometimes swelling and tenderness around the site of the sting. Some nonlethal species may cause more intense pain and greater swelling, with numbness. The bite of the **Deadly Sculptured Scorpion** is said to cause a "strychnine-like reaction," with numbness around the site of the sting, hypertension, salivation and sweating, nausea, semiparalysis of the throat and tongue, convulsions and even death. Muscle spasms and twitching sometimes occur and are believed to be due to the active part of the venom (proteinaceous in nature) displacing calcium where it is bound to the membrane of muscle fibers. Death from scorpion stings is usually due to paralysis of the respiratory system, indicating that the venom affects the central nervous system. The venom is said to be more deadly than that of the rattlesnake, but the amount injected is much less. While many stings of the **Deadly Sculptured Scorpion** are reported annually in the United States, deaths are actually rare. In western Mexico the stings of deadly scorpions (chiefly *Centruroides suffusus, C. noxius, C. infamatus* and *C. limpidus)* are the major cause of death (26).

All children under 12 years of age and elderly persons receiving a scorpion sting of any kind should be put under observation and treatment by a physician or in a hospital as soon as possible. When stings of dangerous species occur on the legs or arms, as they usually do, the recommended first aid treatment is to apply a ligature or tourniquet between the heart and site of the sting, as close to the latter as possible. (a shoelace makes a good ligature, a handkershief does not.) The tourniquet should be released at least every three to five minutes. A piece of ice placed on the site of the sting as soon as possible and held there will help localize the venom, until a mixture of ice and water can be prepared in a suitable container for complete immersion of the limb. The tourniquet should be removed after the limb has been in the iced water for five minutes. In the case of severe stings it is well to leave the limb immersed for at least two hours; it is important to keep the victim warm during this time. Artificial respiration may be advisable if there is great difficulty in breathing. Antivenom is available for the dangerous species of scorpions found in the United States and Mexico, but it is presently manufactured only in Mexico.

Children are especially vulnerable to scorpion stings and account for most of the resultant deaths. People are usually stung when moving rocks, lumber and other materials that have been lying on the ground for some time, or when putting on shoes and clothing in which scorpions are concealed; they are often hidden under boxes in homes and even in beds. Shoes and clothing should be shaken out before putting them on. If one sees or suspects that a scorpion is crawling on an exposed part of the body, it should be brushed off, not swatted.

Centipedes

In review, centipedes belong to the class Chilopoda and differ from other arthropods in having a pair of legs on most body segments. Like insects, and unlike arachnids, they have a pair of antennae or "feelers" on the front of the head; like arachnids, and unlike insects, they never have wings. Centipedes are distinctive in having a pair of venomous claws located at the end of the first pair of trunk appendages and not in the "jaws." The appendages, which lie between the mouth and first pair of walking legs, are four-segmented, the sharp, curved claws being the distal or apical segment. Centipedes are predaceous on insects, but are not aggressive and will bite only in self-defense, when cornered or handled. The small species commonly found in temperate North America are harmless, though they may be able to pierce the skin with their claws. The large tropical species — some are a foot long — and some of the fairly large species found under rocks and in homes in the Southwest can bite severely and are feared. [Other soil inhabiting arthropods often associated with centipedes are the millipedes (class Diplopoda); they differ from centipedes in having a

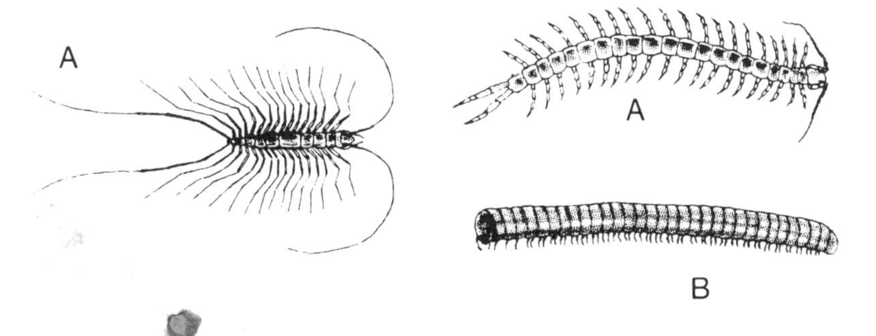

Fig. 66. Centipede (A) and millipede (B).

cylindrical rather than flat body and two pairs of legs on each body segment (also in having the genital openings at the anterior end rather than the last segment). They do not bite, but the larger ones have glands in the segments which secrete a liquid irritating to the human skin.]

The **Western House Centipede** *(Scolopendra heros)* may be recognized by the 21 pairs of legs — one pair on each body segment — and the long anal legs, which are longer than the three terminal body segments. In one of the large outdoor species, *Geophilus californicus,* the legs number from 64 to 67 pairs. Common reactions to centipede bites, besides severe pain and reddening of the skin, are: a rash, swelling and formation of papules and purple patches. The recommended first aid for centipede bites is to dab the area with ammonia and apply a compress of sodium bicarbonate or epsom salts (59).

SUMMARY OF PRECAUTIONS AGAINST ARTHROPODS THAT ATTACK MAN

At Home:

Don't handle unknown insects.

Learn to recognize the dangerous insects of your area.

Don't harbor rodents, wild birds or bats.

Remove vermin from pets regularly.

Cover foods served in unscreened patios.

Clean yourself and your clothing regularly.

Keep your house and closets clean, well swept and screened.

Clear rubbish and brush out of yards.

Wear gloves when disturbing woodpiles and musty corners.

Don't disturb animal and arthropod nests except during the occupants' dormant season — and only when using protection over your face, hands and feet (arthropods may often inhabit the nests of their animal hosts).

Drain pools of stagnant water, keep them disinfected or full of mosquito fish.

Use insecticide (if necessary), but read and heed the label.

Call a professional pest exterminator for help in removing major arthropod infestations.

Outdoors:

Wear protective clothing when hiking through brush.

Use insect repellent as needed.

Wash well after outdoor activity, removing any attached ticks, etc.

Scrub hands thoroughly after skinning game.

Don't handle dead animals found in the open.

Shake out clothes, shoes and packs in the morning when camping and

Shake out bedrolls at night to remove scorpions, spiders and insects.

Check under privy seats for spiders before sitting.

Don't play with wild rabbits or rodents such as rats, squirrels, chipmunks and mice — they may look well, yet carry disease.

First Aid:

In or Near Town —

Take the victim to a doctor immediately if possible, otherwise try to call by phone. Try to capture the arthropod if it is a tick, scorpion or spider for identification by the doctor.

If obvious swelling or discoloration occurs at the wound, keep the patient warm and quiet until a doctor is reached. The venom probably is local rather than general.

If little or no obvious swelling or discoloration occurs at the wound and pain is present, keep the patient very still and the wound chilled until a doctor is reached. The venom probably is general and thus more dangerous. **Time is important.**

If the patient goes into shock, reach the nearest doctor immediately.

Never smash an insect into the wound it has made, and always remove any remaining body parts gently from the wound.

All wounds should be disinfected to avoid secondary infection.

In the Wilds —

Contact your family doctor before long journeys or backpacking for information on shots to take or antivenins to carry with you.

Ticks should be removed with a slow, gentle pull, using your fingers or tweezers.

APPENDICES

APPENDIX 1

THE PLAGUE

Plague, a highly infectious bacterial disease, primarily affects rodents (mice, tree squirrels, ground squirrels, chipmunks, marmots, woodrats, rats, and others) causing varying rates of mortality. It can **affect humans** if they enter infected areas, or if the disease is transferred from wild rodents to rats that live in close association with humans. In California, several cases of plague in humans are reported each year. The victims have usually been bitten by infected fleas from wild rodents.

Recognize the first signs of human plague: fever, nausea, headache, and/or swelling of lymph nodes. If you have been in a plague-infected area, have been around dogs or cats possibly exposed to plague-infected animals, or have any other reason to suspect plague, contact a physician immediately. Tell the physician you may have been exposed to plague.

Notify the local Public Health Agency immediately in all cases where plague is suspected.

How Plague Spreads. — There are four main components associated with the spreading of plague: the plague bacterium *(Yersinia pestis)* that causes the disease, the reservoir rodent that carries it, the flea that transmits it, and the recipient host that receives it.

The bacterium is a rod-shaped organism. Once in the host's body, it causes the disease by reproducing rapidly and generating a powerful toxin that damages tissue. If untreated, the disease is fatal.

The disease reservoir is the place where an infectious agent resides, providing a source for infection. For plague, certain wild rodents serve as the reservoir. Although they have the disease, they may not show obvious symptoms and are usually not harmed by the infection. They are the source of infection for other more susceptible rodents and rabbits that may be more closely associated with humans.

The flea is called the vector of the disease because it carries the bacteria from one host to another. The flea becomes infected with plague and spreads the disease when it bites other host animals. (See pages 87-91 and 175-183.)

Many different species of fleas carry and transmit plague bacteria although some are more efficient than others. Fleas commonly found on domestic rats and ground squirrels are efficient transmitters, whereas those commonly found on dogs and cats are less efficient. However, dogs have been known to carry infected rodent fleas from plague-active areas to campsites and homes.

The recipient host shows typical symptoms when it contracts plague. Humans are very susceptible recipient hosts. Another recipient host, such as the ground squirrel, is usually involved in the transmission of plague to humans.

Plague Transmission. — Most commonly transmitted from ground squirrels to humans via the flea. This might happen as follows: The blood-sucking flea bites an infected ground squirrel and ingests plague bacteria. The bacteria multiply in the flea's forestomach (proventriculus). Eventually the forestomach becomes blocked by the mass of bacteria and the flea is unable to swallow and digest more blood. In an attempt to obtain food, the starving flea bites repeatedly. Each time the flea bites, blood enters the esophagus, is blocked by the bacteria, and surges back to the victim carrying thousands of plague bacteria with it. When infected squirrels die, their infected fleas move to other hosts — another squirrel, a different wild animal, or a human. If it is susceptible to plague, the new host becomes infected and continues the cycle.

A second route of transmission involves pneumonic plague, a form of the disease involving the lungs. The recipient host contracts plague by inhaling airborne bacteria from the breath of infected animals or humans.

Among wild animals, a third source of infection is feeding on other infected animals (predation). Humans can become infected if they eat undercooked, infected meat. Handling diseased animals and inadvertently taking in bacteria from the animal's body fluids through cuts and abrasions can also cause infection in humans.

Animal Hosts. — Worldwide, more than 230 species or subspecies of animals are known to carry plague although only rodents that live in close association with humans (for example, ground squirrels, chipmunks, rats) are considered major threats of plague transmission to humans.

The list of carriers includes several species of ground squirrels, flying squirrels, chipmunks, woodrats, deer mice, voles, shrews, rabbits, pocket gophers, and marmots. Carnivores such as coyotes, bobcats, badgers, gray foxes, skunks, martens, and others can carry plague, but their role in epidemics is considered minor or nil. Most rodents, except the reservoir species such as voles and deer mice, appear to be more susceptible to plague than are canines and other carnivores.

Local outbreak (epizootics) of plague occur periodically among wild animals when the disease moves from the reservoir species to a more susceptible host. A dense population of the susceptible host appears necessary for such an outbreak. Although most susceptible animals that contract the disease die during an epizootic, outbreaks among nocturnal or solitary species often pass unnoticed. Scavengers and predators may also mask signs of an outbreak by removing dead carcasses. For colonial and daytime-active species such as ground squirrels and chipmunks, plague outbreaks are more easily recognized. After an outbreak, bacteria may exist for months in infected fleas in rodent burrows or nests. Susceptible animals may become infected by entering these habitats and being bitten by the infected fleas.

Plague and Man. — Plague is primarliy a disease of small rodents that occasionally strikes humans. Although the threat of rat-borne epidemics still exists, no such outbreaks have occurred in the United States since 1925. Today, the major threat of plague to humans is from wild rodents (particularly ground squirrels) and rabbits. The expanding human population and the development of housing, industry, and agriculture in formerly undeveloped lands, together with life-styles that emphasize outdoor recreation, are bringing people in close contact with certain wild rodent species. This has substantially increased the exposure of humans to plague.

Plague in humans may take three forms: **Bubonic, septicemic,** or **pneumonic.** The most common form, **bubonic plague,** spreads through the body via the lymphatic system. It is characterized by swollen lymph glands, or buboes. Other symptoms may include a high fever of up to 41°C (106°F), headache, and nausea. Initial symptoms usually appear two to six days (but up to 10 days) after infection (for example, from a flea bite).

Septicemic plague, a relatively rare form, enters the bloodstream directly and spreads through the body. The bacteria multiply rapidly, resulting in internal hemorrhaging and swelling of the liver and spleen.

The pneumonic form of plague is centered in the lungs and is highly contagious. Of the three, it is the most dangerous to humans. Plague bacteria develop in the lungs so rapidly that there may not be time for the usual symptoms of bubonic plague to appear. Within one day of infection, there may be signs of a generalized infection, such as a headache and fever. By the second day, coughing, blood in the sputum, and respiratory distress clearly indicate lung involvement. Death may occur two to four days after infection.

Plague is treatable with prompt diagnosis and administration of specific antibiotics. Rapid and proper diagnosis is essential both for the victims and those in contact with them.

Prevention

Discourage and/or reduce rodent populations around home, campsites, and other inhabited areas.

Rats: Control measures for rats include sanitation, exclusion, trapping, and toxic baits. No one technique is totally effective and a combination of all techniques is usually required for satisfactory control. Removing rat harborage and blocking all entrances into buildings (rat-proofing) can greatly reduce rat problems. For limited infestations of rats inside or around buildings, several wooden-base rat traps are usually sufficient to obtain control. For larger infestations, toxic bait may be necessary. Cereal baits, both loose or embedded in paraffin, are widely used. Baits are available in many garden or farm supply stores or from many County Agricultural Commissioners.

Ground squirrels and **chipmunks:** For limited infestations (for example, in gardens or around buildings) control by trapping or fumigating. For infestations in agricultural and similar situations, fumigants or toxic grain baits available from many County Agricultural Commissioners and retail outlets may be required.

Meadow mice (voles): Proper environmental management by elimination of weeds and dense vegetation is important in preventing buildup of populations of meadow mice. For small populations, trapping may be an effective control measure. For larger populations, treated grain baits may be necessary. These baits are available from many County Agricultural Commissioners and some retail outlets.

In plague areas, protect yourself from fleas. Apply insect repellent to socks and trouser cuffs and tuck pants into boots.

Protect pets with flea collars and flea powder, especially if they are allowed to run free in areas where ground squirrels are present. Generally, flea collars alone will not provide complete flea control. When camping in plague areas, pets should be leashed or confined. Better yet, leave them at home.

Be aware of signs of plague in wild animals: Sick, sluggish, or dead animals; sudden, unexplained decreases in or disappearances of an animal (particularly rodent) population; dead animal odors or flies around rodent burrows.

Do not handle sick or dead rodents or other wild animals. Report them to the appropriate authorities (for example, the campground manager, a park or forest ranger, or the public health office). If you must handle dead rodents, use a large plastic bag inverted over your hand and arm to pick up the carcass.

When camping, do not sit or place tent or sleeping bag near rodent burrows. Fleas gather in these areas and wait for a warm-blooded host. Do not feed rodents in campgrounds and picnic areas. Avoid contact with wild animals.

Thoroughly cook all wild game. Use plastic or rubber gloves when skinning or cleaning rodents, rabbits, or carnivores.

Additional information may be found in *PLAGUE, What You Should Know About It,* Leaflet 21233 (1981) by Terrel P. Salmon and W. Paul Gorenzel, published by the Division of Agricultural Sciences, University of California, from which the present verison is abstracted.

APPENDIX 2

FLEAS

During the period "Common Insects of North America," 1972 has been out of print, I have had several requests by various authors and publishers for permission to use some of the illustrations.

The chapter on fleas (pp. 654-660) was used in many classes which discussed the importance of fleas to human health. It seems appropriate to include those illustrations and the accompanying text to provide further information on fleas. The number of animals residing near homes (dogs, cats, etc.) continuously grows and therefore increases the chance of humans becoming infected with diseases transmitted by fleas.

For simplicity, I am using the figure numbers as they originally appeared.

For external morphology, refer to Fig. 35 on page 88.

SIPHONAPTERA: Fleas

As indicated by the name of the order, fleas have sucking mouthparts and no wings. The laterally compressed body, long powerful legs and smooth cuticle with short spine-like hairs directed backwards enable them to move swiftly through the host's hairy or feathery covering. Fleas are parasitic and require the blood of warm-blooded animals to reproduce. The majority of species infest burrowing mammals; some are associated with large carnivores, others with birds and bats. Only man among the primates is a host of fleas. Several species of rat fleas infest homes and buildings which harbor rats, and will bite humans as well. The Oriental rat flea and others transmit plague by biting first an infected rat and then a human; they are vectors of bacteria called *Pasteurella pestis,* the cause of plague, and may themselves die or recover from the infection. Many fleas also carry rickettsiae called *Rickettsia typhi (mooseri),* the cause of "murine or endemic fleaborne typhus," from infected rats to man. [Rickettsiae are considered as intermediate between bacteria and viruses. Like bacteria (and unlike viruses) they can be seen under an ordinary microscope. Like viruses (and unlike bacteria) they do not grow on artificial media and can be propagated only in the host.] Fleaborne typhus occurs in

Europe, Africa, South and Central America, and our southeastern states; it is a milder form of typhus than "epidemic louseborne typhus," which is caused by *Rickettsia prowazeki* and is transmitted by the body louse. [Twenty-some species of fleas found in the U.S. can transmit plague organisms under laboratory conditions. Some are possible vectors of bacteria *(Pasteurella tularensis)* causing tularemia; the common vectors of this organism are ticks and a deer fly.]

Metamorphosis (Fig. 1406) in fleas is complete. The female of most species lays its eggs while on the host, but the eggs usually fall to the floor, ground, or bedding, where they hatch in a few days. The tiny whitish, legless larvae have biting and chewing mouthparts, live in the dirt or dust, and feed on organic matter. They wriggle violently when disturbed, are not normally found on the host. The life cycle in warm, moist climates averages six to seven weeks. Adult fleas can live several weeks without food and are not necessarily eliminated by a short absence from the animals harboring them. Excepting the sticktight fleas, they do not stay with one host but transfer from one individual or one species to another. Fleas do not develop and live in sand for successive generations without feeding on animals; these places are therefore not reservoirs of infestations as some people believe. Throughout the world, more than 1,000 species are known; the authority Karl Jordan believes there are actually twice this number.

(1) Human Flea: *Pulex irritans* (Pulicidae). **Fig. 1407.**

Range: Throughout the world. **Adult:** Light at first changing to dark brown. Preantennal region of head with two long bristles: one below the eye (see Oriental Rat Flea), the other at the base of the maxilla; each thoracic segment with a row of bristles alternately long and short. Genal and pronotal spines absent. It can jump 6 to 8 inches high and 12 to 15 inches horizontally. Adult life may extend to two years or more. The female lays upwards of 500 eggs on animals or floors of houses, barns, or other shelter. **Length:** .04″. **Larva:** Whitish, cylindrical, sparsely covered with hairs. Life cycle from egg to adult varies from 15 to 65 days, depending on temperature, moisture, and food supply. **Host:** Man, dogs, cats, rabbits, coyotes, deer, horses, mules, rodents, and pigs (mainly the last two). [The world-wide occurrence of the "so-called human flea" and its association with man (and not other primates), is believed by

Holland to be through the pig. He notes (1969) that "in New Guinea pigs are kept close to and often inside the household where piglets may be nursed by native women along with their own children."] A possible vector of plague, an intermediate host of the rodent tapeworm *Hymenolepis nana,* which is known to infest man [see (2) below]. *P. simulans* — "long misidentified as *P. irritans"* — is the common flea of coyotes.

(2) Dog Flea: *Ctenocephalides canis.* **Fig. 1406.**

Range: Cosmopolitan. **Adult:** It is easily differentiated from the human flea by the elongated anterior portion of the body, long labial palpi, genal ctenidium (comb) of 7 or 8 spines and pronotal ctenidium of 14 to 16 spines. The female lays about 70 eggs on hairs of the host or on the floor. **Length:** .07″ (male) to .15″ (female). **Larva:** Whitish, elongate, cylindrical, with sparse covering of hairs. Life cycle from egg to adult requires from 27 to 40 days. **Host:** Dogs, cats, man, rats, squirrels, rabbits, poultry. An intermediate host of the dog tapeworm *Dipylidium caninum* and the rodent tapeworm *Hymenolepis nana,* both of which occur in man, especially children. Eggs of the tapeworm are ingested by the flea larva, and the tapeworm completes its development in the adult flea. Dogs are infected by nipping the fleas and pass them on to the children when allowed to lick their faces. The dog flea may also be a vector of plague, and has been known to harbor *Rickettsia typhi,* the cause of murine or endemic typhus.

(3) Cat Flea: *Ctenocephalides felis.* **Fig. 1408.**

Range: Cosmopolitan. **Adult:** Very similar to *C. canis* and often mistaken for it; they were considered the same species for a long time. It may be distinguished from other fleas by the shallow, gradually sloping head, with first two genal spines equal in length; inner side of hind femur has 7 to 10 bristles (there are 10 to 13 in the dog flea). **Length:** .08″ (female); male slightly less. **Host:** Same as dog flea. With *C. canis* (2), it may be a vector of a nematode causing "heartworm" of dogs and occasionally cats [see Diptera (8)]. The cat flea can also transmit murine or endemic typhus.

(4) Oriental Rat Flea: *Xenopsylla cheopis.* **Fig. 1409.**

Range: Cosmopolitan. **Adult:** Like the human flea (1) and chigoe (12), it lacks the genal and pronotal ctenidia displayed by the dog and cat fleas; it has ocular bristle in front of eye (see

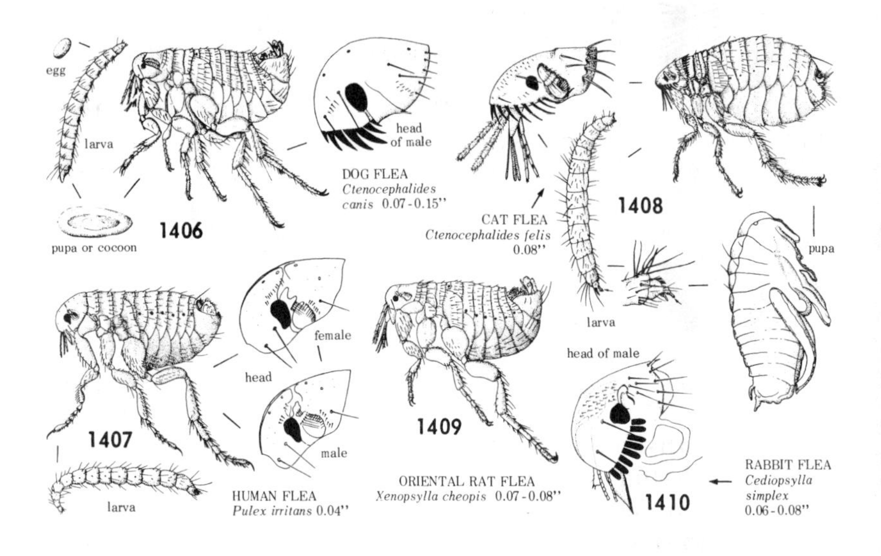

human flea). **Length:** .07-.08″. **Larva:** The larval period lasts 30 days or more. The pupal stage is comparatively long: from 25 to 34 days. Adults live as long as a year. **Host:** Rats and other rodents, man. It is the principal vector of *Pasteurella pestis,* the cause of plague; also a carrier of *Rickettsia typi,* the cause of murine or endimic fleaborne typhus. This is the flea primarily responsible for the bubonic plagues — Black Death — of the Old World during the fourteenth and seventeenth centuries. [A pneumonic form which spreads directly from one person to another usually occurs during epidemics. This form, resembling the Black Death of medieval Europe, caused the death of 60,000 people in Manchuria in 1911; it is believed to have stemmed from the Siberian marmot, which was hunted for furs.] It is an intermediate host of the rodent tapeworms *Hymenolepis diminuta* and *H. nana* [see (2) above], both of which frequently infest children. The Rabbit Flea, *Cediopsylla simplex* (Fig. 1410) — a true parasite of the eastern cottontail *(Sylvilagus floridanus),* also found on the bobcat, red fox, and short-tailed shrew — may be recognized by the angulate head, nearly vertical genal ctenidium of 8 spines on a side, and arrangement of bristles on the head; pronotum has single row of bristles and ctenidium of 6 or 7 spines on a side; it is from .06″ to .08″ long.

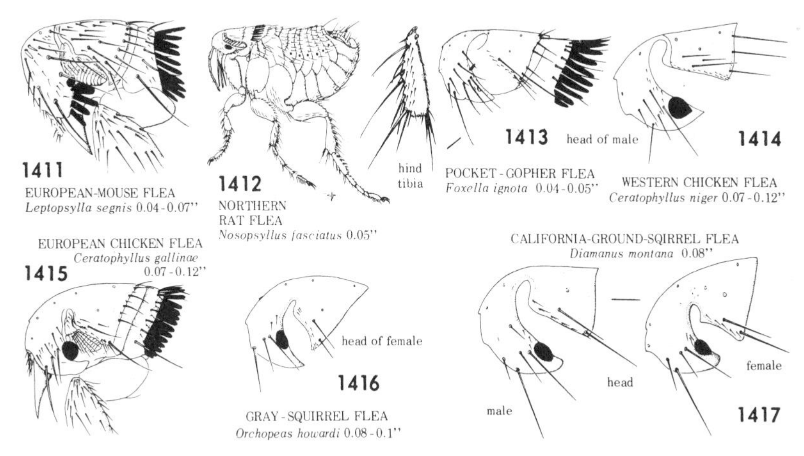

(5) European-Mouse Flea: *Leptopsylla (Ctenopsyllus) segnis* (Ceratophyllidae). **Fig. 1411.**

Range: Widely distributed in Europe and North America, especially the port cities. **Adult:** Structural features of head and pronotum, as shown in the figure, serve to distinguish this flea. **Length:** .07-.07″. **Host:** Rats, mice, men. A true parasite of the European house mouse *(Mus musculus),* now common in North America. It is a "weak vector of plague," capable of transmitting murine or endemic typhus; it seldom bites man, and is considered a negligible factor in outbreaks of plague.

(6) Northern Rat Flea: *Nosopsyllus fasciatus.* **Fig. 1412.**

Range: Cosmopolitan. **Adult:** It may be recognized by the shape of the head, well-developed eyes, pronotal ctenidium of 18 to 20 teeth, lower row of three setae in the male (two in the female) in front of the eye. **Host:** Brown or Norway rat *(Rattus norvegicus),* black or roof rat *(R. rattus),* various house rats, wood rats, mice, other rodents, man. It is not believed to be a significant factor in natural outbreaks of plague. Like the Oriental rat flea (4), it is an intermediate host of *Hymenolepis diminuta,* a common tapeworm of rats and mice which frequently infests children. The Pocket-gopher flea, *Foxella ignota* (Fig. 1413) — a true parasite of the eastern *(Geomys bursarius)* and northern *(Thomomys talpoides)* pocket gophers, also found on white-footed or deer mice, and moles — is distinguished by the bristles on the head and hind tibia, pronotum with row of alternately long and short bristles and ctenidium of 10 or 11 spines; nine geographic variations, or subspecies, have been named.

(7) Western Chicken Flea: *Ceratophyllus niger.* **Fig. 1414.**

Range: Western part of North America. **Adult:** A very dark species, as implied by its name; it may be distinguished from other fleas found in chicken houses by frontal outline of head with sharp in-cut, comparatively large, egg-shaped eyes, and arrangement of bristles in front of the eyes. Pronotal ctenidium consists of 26 to 28 spines. It contacts the host only when it wishes to feed. **Length:** .07″ (male)-.12″ (female). **Host:** Fowl, sparrows, pigeons, man. They are destructive to young poultry during warm summer days if unchecked. The closely related European chicken flea, *C. gallinae* (Fig. 1415), occurs in the New England states, may be identified by the lateral row of 4 to 6 bristles on inner surface of the hind femur.

(8) California-Ground-Squirrel Flea: *Diamanus montana.* **Fig. 1417.**

Range: Washington to California, east to Nevada, Utah, Arizona, Colorado, and New Mexico. **Adult:** Distinguished by arrangement of bristles on the head, as shown in the figures, and by the long spine at the top of the second joint of the hind tarsus extending beyond the third and over the fourth joints; abdominal tergites each with two rows of bristles. **Length:** .08″. **Host:** California (or Beechey) ground squirrel *(Otospermophilus beecheyi);* seldom found on other rodents, usually the only flea found on this squirrel, whose burrows are "teeming with them" during July and August. A moderately efficient vector of plague. [Sylvatic (wild rodent) plague is endemic in the western third of North America. Natural epidemics flare up occasionally among rodents, but only a few outbreaks have involved man.] Franklin's-ground-squirrel flea, *Opisocrostis bruneri* — a large flea, from .12″ to .15″ long — has two rows of bristles on preantennal region of head, a postantennal row of four or five bristles, a single row of alternately stout and weak bristles on the pronotum, two or three rows of bristles on the mesonotum, and about five rows on the metanotum; it ranges from Manitoba to Alberta, Illinois and Wisconsin to Colorado and Montana, is a true parasite of Franklin's ground squirrel *(Citellus franklinii)* and an efficient vector of plague. The Gray-squirrel flea, *(Orchopeas howardi (wickhami)* (Fig. 1416) — from .08″ to .1″ long — may be recognized by the arrangement of bristles on the head; pronotum with single row of bristles and ctenidium of nine spines on a side; it is abundant in the East, a true parasite of the gray squirrel *(Sciurus carolinensis).*

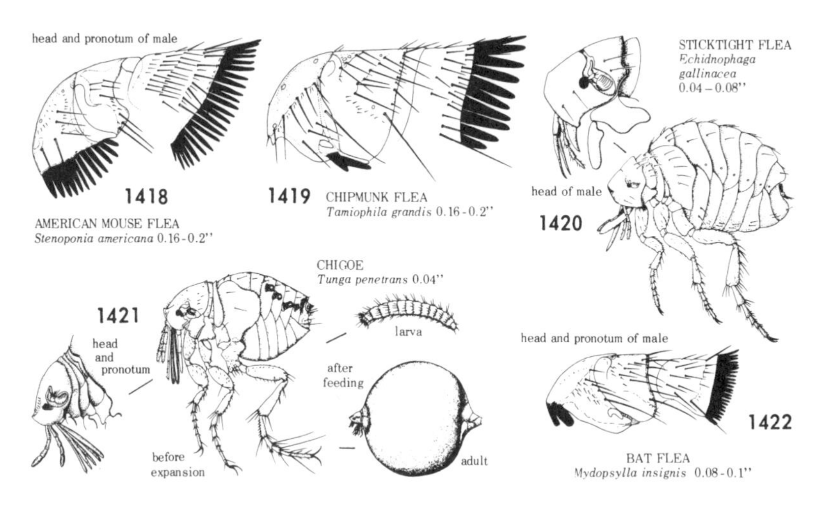

(9) American Mouse Flea: *Stenoponia americana* (Hystrichopsyllidae). **Fig. 1418.**

Range: New Brunswick to Manitoba, south to Virginia, Tennessee, Alabama, west to Montana. **Adult:** Genal ctenidium composed of 13 teeth on each side; pronotal ctenidium consisting of 25 or 26 teeth on each side. **Length:** .16-.2″. A large species, abundant in the fall and winter months. **Host:** Rodents and insectivores; favored host is the white-footed (or deer) mouse *(Peromyscus).*

(10) Chipmunk Flea: *Tamiophila grandis.* **Fig. 1419.**

Range: Massachusetts, New York, Michigan; Nova Scotia to Manitoba. **Adult:** It may be distinguished by the arrangement of bristles on the head and the genal ctenidium of two spines: the first much wider and shorter than the other; pronotum with two rows of bristles and a ctenidium of 10 or 11 spines on each side. In this family the eyes are absent or vestigial. **Length:** .16-.2″. A large species. **Host:** Eastern chipmunk *(Tamias striatus),* its true host; also cottontail rabbit, red squirrel, and weasel.

(11) Sticktight Flea: *Echidnophaga gallinacea* (Tungidae). **Fig. 1420.**

Range: Eastern, southern, and southwestern states; along the Pacific coast into Oregon. **Adult:** Dark brown, almost black. Distinctive features are: head angulate in front, divided by groove, with two bristles in each of pre- and post-antennal regions; eyes heavily pigmented, mandibles deeply serrated and tapering toward the end; single row of bristles on pronotum and mesonotum, no ctenidia; patch of 18 to 20 bristles on inner side of hind coxa. **Length:** .04″ (male), .08″ (female). **Host:** Poultry, birds, dogs, cats, rabbits, ground squirrels, rats, horses, man. They attach themselves in masses on the face, wattles, and earlobes of chickens and turkeys, and remain there, excepting the males which tend to move about at night. They are very damaging to chickens if unchecked, and occasionally get under the skin of poultry handlers. A vector of murine or endemic typhus.

(12) Chigoe: *Tunga penetrans.* **Fig. 1421.**

Also called such names as sand fleas, tique, chigue, pico, bicho, and jiggers (not to be confused with chiggers — mites which cause reddish welts on the human skin). **Range:** Mexico, Central and South America, other tropical regions; southern United States sporadically. **Adult:** It may be distinguished from the sticktight flea (11) by its smaller size and the absence of a patch of bristles on the hind coxa. The female acts like any other flea until it has mated; it then bores under the skin to lay eggs, and remains there. As the eggs develop, the body of the flea becomes enormously distended, forming a sphere which may reach the size of a pea. The eggs are usually extruded from an opening in the skin, and most of them fall to the ground, where they hatch in a few days. **Length:** .04″. **Larva:** Most of the tiny white, hairy larvae develop on the ground like other fleas. Some eggs remain in the wound and the larvae develop in the tissues, causing a sore similar to the lesion caused by blow flies. It normally feeds for six to twelve days, spins a cocoon, pupates, and emerges in about five days. **Host:** Poultry, birds, dogs, cats, man. Badly infested Brewer's blackbirds and English sparrows have been found in Texas. On poultry the fleas are found mostly around the eyes and comb; on dogs and cats they lodge around the ears; they attack man mainly under the toenails and between the toes, causing lesions which may result in tetanus infection or gangrene.

(13) Bat Flea: *Myodopsylla insignis* (Ischnopsyllidae). **Fig. 1422.**

This family is distinguished from the others by the hardened "preoral plates" or flaps which resemble ctenidia and are located near the front on each side of the oral margin. *M. insignis* is the most common and widely distributed of the North

American bat fleas. **Range:** Kentucky to Maine, west to the Rocky Mts. and across Canada. **Adult:** Four or five stout bristles near front of antennae, postantennal bristles in four rows. Pronotal ctenidium with 18 to 21 slender spines on each side. **Length:** .08-.1″.

APPENDIX 3

Rabies

Rabies, or hydrophobia, is caused by a filterable virus — a type of infective agent that can pass through certain filters that retain ordinary bacteria and cannot be seen through ordinary microscopes. The virus is found in the saliva of affected animals. The disease is produced by the bite of a rabid animal or by contact with the saliva of a rabid animal. The bite makes a wound in which the virus in the saliva is deposited.

The disease is primarily one of the dog, although many species, including people, are susceptible to infection.

Rabies has been reported in the cat, cow, horse, mule, sheep, goat, hog, wolf, fox, coyote, hyena, skunk, monkey, deer, antelope, camel, bear, elk, polecat, squirrel, hare, rabbit, rat, mouse, jackal, badger, marmot, woodchuck, porcupine, weasel, hedgehog, mongoose, gopher, raccoon, the owl, hawk, chicken, pigeon, stork, and several species of bats.

When rabies is once controlled in dogs its importance from an economic or public-health standpoint will be greatly reduced. When a disease becomes established in a wild species, however, a serious situation develops, and strenuous efforts must be made to control it in the species affected.

Rabies in foxes became rather wide-spread after 1945 in certain states, and considerable livestock losses occurred when rabid foxes attacked farm animals, particularly cattle.

The number of cases of rabies in all species between 1938 and 1955 was 146,627 — a staggering total. The financial loss to the country runs into large figures and the constant hazards to human beings and animals call for strong national action to control and eventually eradicate the disease.

The period of incubation — the time between exposure to infection and the first appearance of symptoms of the disease —

may be as short as 2 weeks or as long as many months. Most cases develop within 3 months.

All animals or persons bitten do not develop the disease. The proportion of persons who contract the disease after being bitten by rabid dogs and who were not treated has been estimated at 15 percent.

From 35 to 45 percent of the dogs, 40 percent of the horses, 36 percent of the hogs, and 25 to 30 percent of the cattle bitten by rabid animals contract the disease. Whether an individual animal contracts the disease depends in part on the location and size of the wound, the amount of bleeding produced, and other conditions. In general, the nearer the bite is to the central nervous system and the deeper the wound, the greater is the danger of infection. If the hemorrhage resulting from the bite is profuse, the possibility exists that the virus will be washed out of the wound.

After it is deposited in the wound, the virus remains latent for an extremely variable period, which depends on the size, location, and depth of the wound and the amount of virulent saliva introduced. The virus follows the course of the nerves to the spinal cord and along the spinal cord to the brain before the symptoms appear. The period between the bite and the appearance of the first symptom may be 14 to 285 days.

The bite of a dog may be infectious at least 3 days before the dog manifests symptoms of rabies. In one case in the Pasteur Institute at Athens, Greece, infection was found to be present in the saliva 8 days before the dog showed signs of the disease.

A dog can transmit rabies through a bite only if it is rabid. A normal dog cannot transmit the disease. Rabid foxes and other wildlife, through their attacks upon animals, have been responsible for outbreaks of rabies.

From a practical standpoint, there seems to be little danger in consuming milk from even a rabid animal. Milk secretion usually is considerably diminished at the clinical appearance of the disease, so that it is quite unlikely that the animal's milk has been consumed after symptoms have appeared. Nevertheless, the milk from a rabid or suspected cow should be condemned as unfit for consumption.

The same position should be taken with regard to the meat from a rabid or suspected animal. The disposition of farm animals that have been exposed to the bites of a rabid dog but do not show signs of the disease also presents a problem. The

International Rabies Conference in Paris in 1927 made the following recommendation on this point: *"Animals bitten by rabid animals, whether treated or not after the bite, should not be butchered between the eighth day and, at the very least, the end of the third month following the bite."*

In **dogs** the first symptom of rabies may be a change in behavior. The animal may become restless and excitable. A friendly dog may become irritable and snappy, and a dog that ordinarily is less amiable may become friendly. Later it may have a tendency to wander and may disappear for a day or two, returning exhausted and considerably emaciated. The dog may seek dark corners and hide. At times the bark might change in tone.

Later on the dog develops partial paralysis and has difficulty in drinking, although it may make efforts to lap water. It staggers around until complete paralysis sets in. Because the virus attacks the brain and spinal cord and sets up degenerative changes, the symptoms — excitability, convulsions, and paralysis — can be correlated with changes in the central nervous system. The inability to swallow results from paralysis of the muscles of the throat.

The term "hydrophobia" means fear of water, but its application to the disease in dogs appears to be a misnomer, because affected dogs show no fear of water. The use of the term probably has its origin in the fact that affected dogs and human beings have been observed to develop convulsions through their unsuccessful efforts to drink. Even the mere thought of drinking on the part of human beings has been responsible for convulsions, and a dread of water or drinking becomes established in many patients.

In the furious form of the disease, the animal is aggressive. It snaps at objects placed before it and will attack dogs and people. It may attack the bars of a cage with such vigor as to break its teeth. The dog's tendency to roam, its restlessness, and its inclination to bite spread the disease widely.

Rabies may take a number of forms, and the symptoms I described are those of a typical case. Many animals affected with rabies do not exhibit these symptoms. The symptoms may be more or less masked and may perhaps be manifested eventually only by paralysis.

Paralytic symptoms are the outstanding feature in the dumb form of rabies. The dog is not vicious. It has no tendency to bite or roam. It is not excitable. In fact, it may be the opposite.

The outstanding clinical feature of dumb rabies is paralysis of the lower jaw, or dropped jaw. The animal's mouth stays open an inch or more. It can be closed with the hands, but the dog cannot close its own jaw.

Many times this is mistakenly thought to be due to a bone in the throat, and many persons have exposed themselves to the virus by examining the mouth and throat for a bone. Usually a dog with a bone stuck in its mouth or throat makes repeated efforts with its paws or otherwise to remove the bone, but in dumb rabies the animal makes no motion about the head with its paws. An animal with a dropped jaw should be viewed with suspicion. No examination should be made of the throat. The animal should be taken immediately to a veterinarian for diagnosis.

Also in the dumb form of rabies the animal shows evidence of paralysis of the hindquarters and forequarters within a few days. It eventually becomes completely paralyzed and dies.

The course of the disease in both the furious and the dumb forms is usually short, and the animal dies in 3 to 7 days.

Cats that have rabies generally hide under the furniture or in a dark corner. There they may die unobserved in a day or two. As a rule, however, the disease in cats implies danger for human beings. The rabid cat becomes bellicose. From its dark corner it may suddenly attack animals or persons, especially children, jumping up to the face and inflicting severe wounds with its teeth and claws. In the violence of this attack it frequently bites itself.

The rabid cat nearly loses its voice and is able only to mew hoarsely.

Later it loses its appetite, has difficulty in swallowing, becomes emaciated, and succumbs within several days with symptoms of paralysis.

Cattle are susceptible to both furious and dumb rabies. The former is the more common. A sharp distinction cannot always be drawn between the two, however, as the furious type usually merges into the dumb type because of the paralysis that always appears before death. In typical cases of dumb rabies, the paralysis occurs at the beginning of the attack and remains until the animal dies.

The first signs are loss of appetite, stopping of the secretion of milk, great restlessness, anxiety, manifestations of fear, and a change in disposition. There follows in a day or two a stage of excitation or madness, which is indicated by increasing restlessness; loud bellowing, with a peculiar change in the sound of the voice; violent butting with the head and pawing the ground; and an insane tendency to attack other animals, although the desire to bite is not so marked in cattle as in dogs.

About the fourth day the animal usually becomes quieter, and the walk is stiff, unsteady, and swaying — the final paralysis is coming on. Loss of flesh is rapid. Even during the short course of the disease, the animal becomes extremely emaciated. The temperature usually remains about normal or even subnormal. Finally there is complete paralysis of the hindquarters, the animal is unable to rise, and (except for irregular convulsive movements) lies in a comatose condition. It dies usually 4 to 6 days after the first symptoms appear.

The rabid fox, through its attack on cattle, has been responsible for heavy losses in cattle in some areas. In 1953 there were 1,033 cases of rabies in foxes and 1,028 cases in 1954.

The **vampire bat** *(Desmodus rotundus murinus* or *Desmodus rufus)* has been responsible for the spread of rabies to cattle and human beings in parts of Mexico, Central America, and South America, to the warm parts of which the vampire bat is indigenous.

The vampire bat feeds on the blood of animals and man, and an infected bat so transmits rabies. Reports have indicated that in Mexico the vampire bat is moving northward and eastward from the western mountain areas and has been reported in Chihuahua, within 100 miles of the United States border. A survey in southern California in 1952 by the Pan American Sanitary Bureau revealed evidence of the presence of the vampire bat in that area.

Rabies has been found in several insect- and fruit-eating varieties of bats in the United States. Some persons have been attacked by these bats, which, on laboratory examination, were found to be affected with rabies. The first bat found to be affected with rabies was in Florida. It was a Florida Yellow Bat *(Dasypterus florindanus)* and was killed while attacking a boy in Tampa in June 1953. In September 1953, a bat, which was found to be rabid, attacked a woman near Carlisle, Pa. Its species could not be determined because the carcass had been destroyed.

Rabid insect- or fruit-eating bats have since been reported in Texas and Montana. Just where the infection in these bats originated is not known, but the findings indicate that a new problem may be facing those concerned with the control of rabies in the United States.

Many species of wild animals are susceptible to rabies.

Severe outbreaks of rabies among wild animals, especially **coyotes,** appeared in Oregon, California, Nevada, and Idaho in 1950. A rabid coyote bit and caused the loss of 27 steers in one feed lot. Serious outbreaks have been reported since that time.

Outbreaks of rabies in **foxes** occurred in Maine in 1934, in Massachusetts in 1935, and in 1940 in Georgia, South Carolina, and Alabama. The origin of that infection is not known, but of 291 fox heads examined in Georgia, 30.2 percent were affected with rabies. At least 90 head of livestock were reported to have developed the disease from exposure to rabid foxes, and the antirabic treatment was given to 17 persons who were exposed.

In 1954, the red fox and the gray fox were to blame for outbreaks along the Appalachian range from New York to Florida and westward across the Southern States to eastern Texas.

Skunks and **civet cats** have been responsible for outbreaks in Iowa, Missouri, Minnesota, Nebraska, South Dakota, and North Dakota.

Dogs suspected of having rabies or dogs that have bitten people should be held in strict quarantine in suitable strong, tight, escapeproof quarters for 2 weeks in order that a correct diagnosis may be made. If a rabid dog is not properly confined, it may work its way out and disappear and die later at a distant point without having been identified. That would complicate the handling of a case of a person bitten by the dog. The fact that the dog had escaped would in itself be some indication that it might have been rabid, but, without proof, an uncertain situation develops.

Outbreaks of rabies in dogs continually occur in certain sections of the United States. In some of the outbreaks, vigorous steps are taken to bring the disease under control and to eradicate it. In other areas little is done to reach that goal, primarily because of lack of organization and funds. As long as rabies exists to any appreciable extent in the dog population in any locality, areas that are free of the disease are menaced. Not only does a rabid dog travel far; very likely dogs in the incubative stage of the disease are transported in automobiles, and so may spread the disease from an infected area to a distant clean one. Energetic measures will bring rabies under control and lead to its eradication in an area, but it can be introduced readily again from an infected locality.

Outbreaks of rabies among wildlife usually occur in areas where the various species have reached a high density of population. The procedures then include the reduction of the numbers of animals in the area through trapping, shooting, gassing dens, and poisoning. Those procedures should be handled by trained specialists and should be undertaken in cooperation with the Fish and Wildlife Service and the corresponding agencies in the states. As in the control of rabies in dogs, the public should be informed about the need for the program.

This Appendix is an extract from an article authored by Dr. W.H. Schoening and originally published in the U.S.D.A. yearbook for 1956.

Rabies is *not* caused by insects. However, it seems useful to add these few pages to this book since it deals with the out-of-doors.

GLOSSARY

Abdomen: The posterior or hind part of the body; in insects it has eleven or twelve segments, not all of which are visible.

Allergy: Extreme sensitivity (hypersensitivity) to a particular substance, in amounts not causing a reaction in most persons.

Anophelenes: Mosquitoes of the genus *Anopheles* (the malaria mosquitoes).

Antenna (pl. Antennae): Paired, segmented "feelers" on front of head, as found in insects, centipedes, and crustaceans.

Antibiotic: A chemical substance formed by certain microorganisms — fungi and others — that destroys bacteria and other microorganisms.

Antibody: A protein produced in the body when a foreign substance (an antigen) is introduced into the tissues. The two substances combine chemically; antibodies tend to be very specific. *See* ANTIGEN.

Antigen: A substance that stimulates formation of an antibody *(which see).*

Antihistamine: A drug used to counteract the action of histamine in allergic reactions. *See* HISTAMINE.

Antitoxin: A substance formed by the body to counteract a toxin, and contained in the blood serum. *See* TOXIN and SERUM.

Antivenin: Antitoxin produced by injection of the venom, or serum containing the antitoxin *(which see).*

Anus: Opening at the lower end of the digestive tract. *adj.* ANAL, sometimes used to designate the posterior part of the body.

Apex (pl. Apexes or Apices): Part farthest from the base *(which see).*

Aphrodisiac: *n.* Drug causing sexual stimulation; *adj.* having this effect.

Appendage: Any subordinate part attached to the body; e.g. tail, cerci, legs, claws, palps, antennae.

Arachnids: Animals belonging to the class Arachnida; spiders, daddylonglegs, mites, ticks, scorpions, whipscorpions, pseudoscorpions, solpugids or windscorpions, king or horseshoe crabs.

Areole (or Areola): The dark ring around a raised area such as a nipple.

Arthropod: An animal belonging to the phylum Arthropoda. *See* PHYLUM.

Avian: Pertaining to birds, a word derived from the class name Aves.

Bacteria (sing. Bacterium): Single-celled microorganisms, visible under an ordinary microscope; they can be cultured (grown) on artificial media. Some cause disease, others are important in digestive and fermenting processes and in providing food for plants through breakdown of soil components and fixation of nitrogen. *See* RICKETTSIAE and VIRUS.

Base: The bottom on which a thing stands or rests; part nearest the point of attachment to the body (in insects base of abdomen is part nearest the thorax, base of thorax is part nearest the abdomen). *See* APEX.

Bite: The wound made by the mouthparts of insects and other arthropods; the effect may be immediate or delayed. *See* STING.

Bovine: *n.* Cow or ox; *adj.* pertaining to cow or ox.

Brood: All the individuals hatching from eggs laid about the same time by a given female.

Cannibalistic: Given to eating others of the same kind.

Canthariasis: Disease due the presence of beetle larvae in the body.

Capitulum: The "head" region of mites and ticks projecting from the underside or anterior end of the body.

Carbohydrate: Organic compounds of carbon, hydrogen, and oxygen forming the starches, sugars and celluloses. Carbohydrates are present in plants and in much lesser amount in animals, and are essential as nutrients to all organisms.

Carnivorous: Flesh-eating. *See* HERBIVOROUS.

Caterpillar: Young larva of moths and butterflies, often called "worms".

Cephalothorax: Head-thorax combined as one; e.g. arachnids, crustaceans.

Cerci (sing. Cercus): Paired appendages at the tip of the abdomen, serve a sensory function and sometimes used for clasping.

Chelicera (pl. Chelicerae): Paired cutting organ forming part of the capitulum in mites and ticks; two-segmented "jaws" of spiders, the apical segment being the claws or fangs; first head appendages.

Class: A unit of classification comprising a group of orders. *See* PHYLUM.

Clypeus: Lower part of the head below the front.

Compound Eye: Large lateral pair of eyes of insects and crustaceans, with several of many facets or sides with associated light-sensitive cells and refractive system. An eye of a bee has many thousands of facets with corresponding "cones" and retinal rods which together form images.

Concave: Surface curving inward. *See* CONVEX.

Conjuctivitis: Infectious inflammation of the conjunctiva or mucous membrane lining inner surface of the eyelids.

Convex: Surface curving outward. *See* CONCAVE.

Copulation: The act of mating.

Corium: Hardened part of the wing in the Hemiptera or true bugs.

Coronary Trombosis: Blood clot obstructing coronary artery.

Coxa (pl. Coxae): First or basal segment of the leg.

Crustaceans: Animals belonging to the class Crustacea, usually aquatic and gill-breathing, with hard shell, two pairs of antennae and one pair of mandibles; e.g. crabs, shrimps, sand hoppers, barnacles, sowbugs.

Cutaneous: Having to do with the skin.

Cuticle: Non-cellular outer covering of the skin. *See* EPIDERMIS.

Deciduous: Usually used in reference to trees that shed their leaves in the fall.

Dermatitis: An inflammation or eruption of the skin.

Desensitize: Make insensitive (non-allergic) to venom or other substance; removing antibodies (proteins produced in the body in response to an antigen or foreign substance) from sensitized cells.

Dimorphic: Having two forms of individuals in the same species.

Disease: Any abnormal condition of the body causing illness; often caused by invasion of foreign microorganisms.

Distal: Part farthest from body or point of attachment.

Diuretic: *n.* Drug causing increased flow of urine; *adj.* having this effect.

Dorsal: On or near the upper surface. *See* VENTRAL.

Drone: The male bee, more especially in a honey bee colony; his sole function is to mate with a queen, after which he dies.

Elytra (sing. Elytron): The hardened forewing of beetles; they serve as a cover for the membranous hindwings when not flying.

Embolus (pl. Emboli): Copulatory organ of the male spider. *See* PEDIPALPS.

Endemic: Continually occuring in a region; a disease is said to be endemic when normally present in an area but at a low level. *See* EPIDEMIC.

Engorged: Glutted or filled to capacity, with abdomen visibly distended.

Enzymes: Organic compounds that are found in animal and plant cells and act as catalysts, causing chemical changes to take place in other compounds having to do with metabolism and other processes necessary to sustain the organism.

Epidemic: Rapid spread of the disease or contagion. *See* ENDEMIC.

Epidermis: Outer layer of cells of the skin underlying and secreting the cuticle (*which see*).

Erythema: Abnormal redness of the skin due to congestion of capillaries.

Excrement: Waste matter eliminated from the anus; feces.

Excretion: "Getting rid of products of metabolism (either by storing them in insoluble form, or by removing them from the body)."

Family: As used in biological classification, a group having certain similarities; a family comprises a group of related genera. The name begins with a capital letter and ends in "idae"; it is converted to a common noun or adjective by dropping the "ae" and not capitalizing the first letter. *See* GENUS.

Feces: Excrement (*which see*).

Femur (pl. Femora): Third segment of the leg, usually the stoutest.

Filariasis: Disease caused by filarial worms (microscopic round worms belonging to the class Nematoda) which invade the lymphatic tissues and are transmitted by mosquitoes.

Genitalia: The reproductive or sexual organs, especially the external parts.

Genus (pl. Genera): A group of species considered to be closely related structurally or otherwise. A scientific name when complete usually consists of the genus and species (or sometimes a third name, the subspecies). A genus name may not be duplicated in the animal kingdom but a species name may be used again in a different genus. The genus and species are commonly written in *italics* or underlined, the genus only beginning with a capital letter. *See* SPECIES.

Gravid: Carrying eggs or young; pregnant. (Literally, it means *heavy.*)

Halteres: The short knob-like structures on the thorax of flies; they take the place of the second pair of wings found in other insects.

Herbivorous: Feeding on plants only. *See* CARNIVOROUS.

Hibernate: To enter into a state of dormancy during winter. Many vertebrates (many mammals, most reptiles and amphibians) and invertebrates hibernate. This period of suspended animation in insects is also called DIAPAUSE, which also occurs under circumstances not associated with the seasons. A similar state occurs with some animals in hot or dry periods and is called AESTIVATION.

Histamine: An organic substance released from the tissues of an allergic reaction to injury, causing dialation of local bloodvessels.

Homopterous: Belonging to the order Homoptera. *n.* HOMOPTERAN.

Hornet: A vespid wasp, usually brown or black with yellow or white bands or other markings. It usually nests in hollow trees or stumps.

Host: The plant or animal on which an insect feeds and develops.

Hypersensitive: Being extremely sensitive to a substance, reacting to amounts which are unnoticed by normal individuals.

Hypodermis: Epidermis (*which see*).

Hypostome: The "beak" or piercing-sucking organ of mites and ticks. *See* CAPITULUM.

Icheneumon: A wasp of the family Ichneumonidae; also called ichneumonid. A very large group, parasitic (as larvae) on other insects.

Immature Forms: Stages not sexually mature or capable of reproduction; the young (nymphs or larvae). *See* LARVA and NYMPH.

Incision: A cut into the body tissue.

Infection: Disease caused by microorganisms.

Infestation: The presence of an organism in very large numbers.

Ingestion: The act of eating or taking in food.

Insect: An invertebrate animal belonging to the class Insecta or Hexapoda (meaning six-legged). An insect has *three* body parts (head, thorax, and abdomen), *three* pairs of legs, and usually wings in the adult stage, and a pair of antennae.

Invertebrate: An animal without a backbone; arthropods belong to this group, the body having an *outer* supporting structure or *exoskeleton*.

Labellum: Tip of the mouthparts of mosquitoes and certain other flies.

Labium: The upperlip, forming the roof of the mouth, in chewing insects or other arthropods. *See* MANDIBLES.

Larva (pl. Larvae): The young of animals having a complete metamorphosis, as in insects, frogs and toads; the stage between egg and pupa in insects, being the growing or developmental stage; the tadpole stage in frogs and toads. Second stage (between egg and nymph) in mites and ticks. Unlike nymphs, larvae are markedly different from adults. *See* METAMOPHOSIS, NYMPH.

Lateral: Toward or related to the side.

Lesion: An injury to the body tissue.

Lethal: Deadly.

Longitudinal: Running lengthwise. *See* TRANSVERSE.

Malphigian Tubules: Excretory glands "opening into anterior part of hind gut of insects, arachnids, and myriapods."

Mammal: Any vertebrate animal in which the female feeds its young by means of milk-secreting (mammary) glands; a member of the class Mammalia.

Mandibles: The first or upper pair of jaws in chewing and biting insects; they may be toothed for chewing or pointed and hollow or grooved for biting and sucking. *See* MAXILLA.

Maxilla (pl. Maxillae): The second or lower pair of jaws of chewing insects and other arthropods.

Median: Being located in the middle, or pertaining to that region.

Metabolism: Breaking down foods and other organic compounds into simple ones, those liberating energy for building up these substances into the complex compounds needed to sustain the body.

Metamorphosis: The transformation from egg to adult. In insects it is said to be complete when four stages occur: egg, larva, pupa, and adult (with the pupa remaining inactive); it is incomplete when a nymphal stage occurs. *See* LARVA and NYMPH.

Microorganism: An organism visible only under microscope, such as bacteria, rickettsiae, protozoa, fungi, and viruses.

Monomorphic: Having only one form. *See* DIMORPHIC and POLYMORPHIC.

Mucous Membrane: The mucous-secreting lining or membrane covering body cavities having openings connecting to the air, such as the intestines and anus, lungs (including mouth and nose), eyes, eyelids.

Myiasis: "Disease resulting from the presence of maggots of flies in or on the living body of man or other animal."

Necrotic: Causing death or decay of tissue. *n.* NECROSIS.

Nectar: A sweet secretion of many flowers, used by bees in making honey.

Neotropical: South America, Central America (all of Mexico), and the Antilles.

Neurotoxic: Damaging to the nervous system. *n.* NEUROTOXIN.

Nodule: Small knot; rounded node or lump. *adj.* NODULAR.

Northwest (Pacific Northwest): Oregon, Washington, British Columbia.

Nymph: Young or immature insects having "incomplete metamorphosis"; stage between larva and adult in mites and ticks.

Ocellus (pl. Ocelli): Simple eyes, as found in insect larvae and spiders.

Omnivorous: Eating both plant and animal food.

Order: A unit of biological classification composed of families having certain similarities of structure; in insects it is usually based on the wings; e.g. Diptera (meaning two wings). *See* CLASS.

Ovipositor: Egg-laying instrument of the female insect, usually located at the end of the abdomen; it reaches its highest development in the parasitic wasps in which it resembles a hypodermic needle.

Palpus (pl. Palps or Palpi): Paired, segmented appendages attached to the insect's mouth. MAXILLARY PALPS are attached to the maxillae, LABIAL PALPS (shorter of the two) are attached to the labium. They are believed to be sensory and have been shown to be organs of smell in some insects.

Papules: Small elevations of the skin (or pimples) which are usually inflamed.

Parasite: An animal or plant that lives at the expense of another, either within or on the host. Parasitic insects which live at the expense of other insects remain on *one* host as larvae, and eventually consume it; the adults are free-living and usually feed on the nectar of flowers. Insects such as lice, fleas, louse flies, and bed bugs, which live at the expense of higher forms of animals, are usually parasitic as adults and free-living part of the time (if only to find or change hosts).

Pedipalps: "Second head appendages of arachnids", sensory in function; in scorpions they have pincers and are used for grasping, in male spiders they are modified as a copulatory organ. *See* EMBOLUS, CHELICERA.

Petiole: Waist, stem or stalk.

Phylum (pl. Phyla): A broad unit of biological classification comprising related classes. Insects (class Insecta), centipedes and millipedes (together called Myriapoda), crustaceans (Crustacea), arachnids (Arachnida) and other smaller groups comprise the phylum Arthropoda (meaning joint-legged). The various phyla are divided into classes, each class into orders, each order into families, each family into genera, each genus into species.

Plumose: Plume-like; feathery, with long processes on each side.

Pollen: The powdery male sex cells found on the stems of flowers, and used by bees and certain other insects as food. *v.* POLLINATE.

Pollen Basket: A smooth concavity fringed with hairs, on the outside of the tibiae of the hind legs of certain bees. It is used for carrying pollen back to the hive for feeding of the young.

Pollen Brush: Rows of bristles on the inside of the basal segment of the tarsus of the hind leg of certain bees. It is used to gather pollen, which is transferred to the pollen basket on the opposite leg.

Polymorphic: Having several forms. *See* DIMORPHIC and MONOMORPHIC.

Predator: An animal that preys on other animals. *adj.* PREDACEOUS or PREDATORY.

Prehensile: Adapted for grasping, usually by winding an appendage around the object; e.g. monkey's or opossum's tail, the antennae of certain insects.

Primate: A member of the order of Primates; includes man, apes, monkeys, and lemurs. The class is Mammalia, the phylum Vertebrata.

Proboscis (pl. Probosces): Extended mouthparts; beak. *See* ROSTRUM.

Progeny: Issue or offspring.

Proleg: Short, fleshy projections on the underside of the *abdomen* of caterpillars, used in walking and clinging to plants; they supplement the true legs which are attached to the *thorax.* Prolegs are lost in the pupal stage during transition from larva to adult.

Proprietary (drug or ointment): One sold under patent, trade-mark, or copyright.

Protein: One of the main components of plant and animal tissues. Proteins are highly complex combinations of amino acids and contain carbon, hydrogen, oxygen, nitrogen, and sometimes sulphur. They are essential as nutrients in sustaining all forms of life and in maintaining the reproductive process. They are also a component of insect venoms and involved in allergic reaction to them.

Prothorax: *See* THORAX.

Protozoa: Single-celled animals, usually microscopic in size, belonging to the phylum Protozoa. They differ from bacteria in having at least one well-defined nucleus. While some are parasitic and important medically, the group as a whole is very important in the economy of nature.

Psychodid: Flies belonging to the family Psychodidae, commonly called moth flies; includes non-biting drain flies and blood-sucking sand flies.

Pupa: The stage of development in insects between the larva and adult; this is a "resting" stage in which the larva transforms into an adult. *See* METAMORPHOSIS.

Pustule: A pimple containing pus.

Queen: The female reproductive bee, ant, paper-nest wasp, and termite colonies. After mating, she founds a new colony or takes over from another queen. When a colony is established, she is fed by the workers and confines herself to egg-laying.

Reduviid: A bug (order Hemiptera) belonging to the family Reduviidae; commonly called assassin bugs, and mostly predaceous; some are blood-sucking and attack man.

Regurgitate: To bring partly digested food up from the stomach and back to the mouth.

Repellent: A substance used to repel pests or prevent them from biting.

Respiratory: Having to do with breathing, or the process of taking in oxygen and giving off carbon dioxide and other products of oxidation. *n.* RESPIRATION.

Rickettsia (pl. Rickettsiae): Single-celled microorganisms similar to bacteria, except that they cannot be grown on artificial media, as bacteria can. They can be seen under an ordinary microscope, and are thought to be intermediate between bacteria and viruses.

Rostrum: Beak or snout; extension of mouthparts or head. *See* PROBOSCIS.

Royal Jelly: A special food secreted by worker bees and fed to young larvae and adult queens.

Salivary Glands: Glands usually opening into the mouth that secrete saliva — a digestive fluid and sometimes an irritant.

Salivation: Excessive flow of saliva.

Scavenger: An animal that feeds on dead or decaying animal or vegetable matter.

Sclerite: Hardened plates or portions of the insect body.

Scutellum: "A somewhat triangular or crescentric sclerite just behind the mesonotum (top of middle thoracic segment)".

Scutum: Shield behind the capitulum, covering front part of body in female hard ticks and the entire body in male hard ticks.

Sebaceous: Skin glands secreting a fatty substance (sebum); they usually open into a hair follicle.

Semen: Fluid secreted by the male reproductive organs (testes) and containing the sperm.

Sensitized: To be made highly sensitive to a serum by repeated injections.

Serum: The clear fluid that separates from the blood corpuscles and contains the antitoxins of a specific disease in animals immunized against it. *See* ANTITOXIN.

Seta (pl Setae): Slender hollow hairs, generally believed to be sensory in function.

Social: Living in communities with divisions of labor and castes; e.g. honey bees, bumble bees, paper wasps, ants, and termites.

Solitary: Not living in communities with division of labor.

Species (sing. and pl.): A group of individuals isolated from others reproductively, mating freely and bearing fertile offsprings. *See* GENUS.

Southwestern (the Southwest): New Mexico, Arizona, southern California, western Colorado, southern Utah and Nevada, western Texas and the Oklahoma panhandle. This is an arid region mostly, and much of it desert, extending into northwestern Mexico.

Sperm or Spermatozoon (pl. Spermatozoa): Male reproductive cell.

Sphecid: Wasps belonging to the family Sphecididae; the so-called sand wasps and mud-daubers, some of which have a long, narrow petiole and are also referred to as thread-waisted wasps; they are solitary.

Spirochetes (Spirochaetes): Elongated, spirally-twisted microorganisms, generally classified with bacteria. Some are free-living, others are parasitic and cause disease in man; e.g. relapsing fever, syphilis.

Sting: A wound — with sharp, sudden pain and burning or smarting sensation — made by the stinger of bees, wasps, ants, and scorpions, or by the prick of nettles. *See* STINGER and BITE.

Stinger: Sharp needle-like weapon at the end of the abdomen in female bees, wasps and ants, and in scorpions; it is used defensively.

Subaculear spine: Blunt thorn at the base of the stinger in the dangerous species of scorpions.

Subcutaneous: Under the skin.

Stylets: Long, slender, sharp processes.

Systemic: Affecting all of the body.

Tabanids: Flies belonging to the family Tabanidae, commonly known as deer flies and horse flies; fierce biters of cattle, horses, and man.

Tarsus (pl. Tarsi): Terminal segment of leg, or the "foot"; it consists of one to five segments, and bear claws.

Telson: The bulbous terminal segment of the scorpion's tail containing two poisonous glands and ending in a sharp stinger.

Testes: Male gonads, where the reproductive cells develop.

Thorax: The middle part or division of the insect body. It always bears the legs and wings, and has three segments: pro-, meso-, and metathorax.

Tibia (pl. Tibiae): Fourth segment of the leg (starting at the base or body) of insects; the fifth segment in the case of spiders.

Toxin: A poison; poisonous substance produced by some microorganisms that cause disease; the proteinaceous substance produced by certain insects and by poisonous spiders, scorpions and snakes, which causes allergic reactions to persons bitten or stung. *adj.* TOXIC.

Transovarial Transmission: Transmission of microorganisms causing disease from one generation to another by way of the egg.

Transverse: Running crosswise or from side to side. *See* LONGITUDINAL.

Trochanter: Second segment of the leg, sometimes divided or fused with femur.

Trypanosomes: Microorganisms of the genus *Trypanosoma* (Protozoa). They are flagellate (having tails), are "parasitic in the blood of vertebrates (including man) and the gut of tsetse flies and other insects, which transmit them to vertebrates". Trypanosomes cause disease in horses and cattle and sleeping sickness in man.

Tubercle: A small rounded projection, or a button-like process.

Tubule: A small tubular process as found on animals.

Urticating: Stinging caused by nettles or poisonous hairs.

Vector: Carrier; an animal that transmits disease organisms from one host to another.

Venom: Poisonous substance injected by some animals when they bite or sting and which causes an allergic reaction in the victim *adj.* VENOMOUS. *See* TOXIN.

Ventral: The underside of the body. *See* DORSAL.

Vertebrate: An animal having a backbone; a member of the phylum Vertebrata.

Vesicant: *n.* A blistering agent; *adj.* causing blisters.

Vespid: A wasp belonging to the family Vespidae; paper-nest wasps (hornets, yellow jackets, and polistes wasps), which are social in behavior, and the potter and mason wasps, which are solitary.

Virus (pl. Viruses): Microorganisms causing disease and visible only under the electron microscope. They multiply in living cells and cannot be cultured on artificial media. *See* BACTERIA and RICKETTSIA.

Wasp: An insect belonging to the order Hymenoptera. Wasps include hornets, yellowjackets, and paper wasps, which are predaceous and social in behavior, and many kinds of burrowing wasps, mud daubers, mason and potter wasps, which are predaceous and solitary. Many of these wasps can sting severely. Parasitic wasps comprise a very large proportion of the total wasps but not many of them sting man. Most of them are very small and go about their business unnoticed. The females lay their eggs in or on the host insect; the larvae develop here and usually consume the host gradually.

Welt: A raised ridge on the skin.

Western (the West): From the Pacific coast to the Rocky Mts.: Alaska, British Columbia, Alberta, Washington, Oregon, California, Arizona, New Mexico, Utah, Nevada, Colorado, Wyoming, Idaho, and Montana. (It is usually considered to be west of the 100th meridian, which would extend the region, as we define it, east of the Rocky Mts. to western Manitoba and about the middle of North Dakota, South Dakota, Nebraska, Kansas, Oklahoma panhandle, and Texas.)

Wheal: A small elevation of the skin, usually from the bite of an insect, with itching; pustule or pimple.

Worker: Infertile female of the social insects; workers do all the work once the colony is established: gathering food, caring for the young and queen, guarding and cleaning the nest. While they do not mate, worker ants sometimes lay eggs; this is true of some paper wasps also. When necessary to replace a queen in a honey bee colony, workers can make one of their own kind into a queen by feeding her royal jelly.

Yellowjacket: A vespid wasp, usually black and yellow or black and white in color; they build globular paper nests suspended from trees or other overhang, or in bushes and in the ground.

BIBLIOGRAPHY

1. **Ainslee, C.N.** 1910. "Papers on Cereal and Forage Insects: The New Mexico Range Caterpillar", U.S. Dept. Agr., Bur. Ent., Bul. 85, Part V.
2. **Aldrich, J.M.** 1915. "The Dipterous Genus *Symphoromyia* in North America", U.S. Natl. Mus. Proc., 49:113-142.
3. **Baker, E.W.** et al. 1965. A Manual of Parasitic Mites of Medical or Economic Importance, New York: Natl. Pest Control Assoc. Tech. Pub.
4. **Baerg, W.J.** 1961. Scorpions: Biology and Effect of Their Venom, Bul. 649, Agr. Exp. Sta., Univ. of Arkansas, Fayetteville.
5. **Bequaert, J.C.** 1955. "The Hippoboscidae or Louse Flies (Diptera) of Mammals and Birds", Ent. Amer., 35:386-410.
6. ———— 1957. "The Hippoboscidae or Louse Flies (Diptera) of Mammals and Birds", *Ibid.,* 36:507-519.
7. **Bishopp, F.C.** and **Philip, C.B.** 1952. "Carriers of Human Diseases", in Insects: Yearbook of Agriculture 1952, Washington, D.C.: Govt. Printing Office.
8. **Blum, M.S.** and **Callahan, P.S.** 1963. "The Venom and Poison Glands of *Pseudomyrmex pallidus* (F. Smith)", Psyche, 70(2):69-74.
9. **Boch, R.** and **Shearer, D.A.** 1965. "Attracting Honey Bees to Crops Which Require Pollination", American Bee Jour., 105(5):166-167.
10. ———— 1965. "Alarm in the Beehive", *Ibid.,* 105(6):206-207.
11. **Brennan, J.M.** 1935. "The Pagoniinae of Nearctic America (Tabanidae)", Univ. of Kansas Sci. Bul., 22:249-401.
12. **Brooks, A.R.** 1967. Aquatic and Semiaquatic Heteroptera of Alberta, Saskatchewan, and Manitoba, Mem. Ent. Soc. of Canada No. 51.
13. **Buren, W.F.** 1970. "Revisionary Studies on the Taxonomy of the Imported Fire Ants", Georgia Ent. Soc. Jour., 7(1):1-26.
14. **Cameron, A.E.,** 1926. "Bionomics of the Tabanidae of the Canadian Prairie Provinces", Bul. Ent. Res., 17:142.
15. **Carpenter, S.J.** and **La Casse, W.J.** 1955. The Mosquitoes of North America (North of Mexico)", Los Angeles: Univ. of Calif. Press.
16. **Cazier, Mont A.** and **Mortenson, Martin A.** 1964. "Bionomical Observations on Tarantula-Hawks and Their Prey (Hymenoptera: Pompilidae: *Pepsis*)", Ann. Ent. Soc. Amer. 57:533-541.
17. **Chandler, A.C.** and **Read, C.P.** 1961. Introduction to Parasitology with Special Reference to Man, New York: John Wiley & Sons.
18. **Coates, Col. J.D., Jr.** and **Hoff, E.C.,** Eds. 1964. Preventive Medicine in World War II (Vol. VII): Communicable Diseases — Arthropodborne Disease Other Than Malaria, Office of the Surgeon General, Dept. of the Army, Washington, D.C.

19. **Cole, Frank R.** and **Schlinger, Evert I.** 1969. The Flies of Western North America, Berkeley: Univ. of Calif. Press.
20. **Cooke, J.A.** 1972. "Stinging Hairs: A Tarantula's Defense", Fauna, July-Aug.:6-8.
21. **Cooley, R.A.** and **Kohls, G.M.** 1944. "The Argasidae of North America, Central America and Cuba", Amer. Mid. Nat. Monograph No. 1.
22. ———1943. "*Ixodes californicus* Banks, 1904, *Ixodes pacificus* n. sp., and *Ixodes conepati* n. sp.", Pan-Pacific Ent., 11(4):139.
23. **Crevecoeur, Hector St. Jean de.** 1782. Letters from an American Farmer, New York: E.P. Dutton, Everyman Paperback, 1957.
24. **Cummings, Carl.** 1933. "The Giant Water Bugs (Belostomatidae: Hemiptera)", Univ. of Kansas Sci. Bul., 21(2):197-220.
25. **Duncan, Carl D.** 1939. A Contribution to the Biology of the North American Vespine Wasps, Stanford Univ., Calif.: Stanford Univ. Press.
26. **Ennik, Franklin.** 1972. "A Short Review of Scorpion Biology, Management, of Stings, and Control", Calif. Vector Views, 19(10):69-80.
27. **Essig, E.O.** 1926. Insects of Western North America, New York: The Macmillian Co.
28. **Evans, H.E.** 1963. Wasp Farm, Garden City, N.Y.: Natural History Press.
29. **Ewing, H.E.** 1928. "The Scorpions of the Western Part of the United States, with Notes on Those Occurring in Northern Mexico",U.S. Natl. Mus. Proc. No. 2730, Art. 9.
30. **Ferguson, D.C.** 1972. The Moths of America North of Mexico, Fascicle 20.2B: Bombycoides, London: E.W. Classey, Ltd. (Ent. Reprint Spec., Los Angeles, Calif.).
31. **Frazier, Claude A., M.D.** 1968. Diagnosis and Treatment of Insect Bites, Clinical Symposia, CIBA Pharmaceutical Co., Summit, N.J.
32. ———1969. Insect Allergy: Allergic and Toxic Reactions to Insects and Other Arthropods, St. Louis, Mo.: Warren H. Green.
33. **Fronk, W.D.** 1963. Increasing Alkali Bees for Pollination, Mimeo Cir. No. 184, Agr. Exp. Sta., Univ. of Wyoming, Laramie.
34. **Gaul, A.T.** 1941. "Experiments in Housing Vespine Colonies, with Notes on the Homing and Toleration Instincts of Certain Species", Psyche, 48:16-19.
35. **Gertsch, W.J.** 1949. American Spiders, New York: D. Van Nostrand.
36. **Gordon, John E.,** Ed. 1965. Control of Communicable Diseases in Man, New York: Amer. Public Health Assoc.
37. **Gorham, J.R.** 1970. The Brown Recluse, Public Health Serv. Pub. No. 2062, U.S. Dept. Health, Education, and Welfare, Washington, D.C.
38. **Gould, D.J.** 1956. The Larval Trombiculid Mites of California, (Acarina: Trombiculidae), Univ. of Calif. Publs. Ent., Vol. 11.
39. **Gregson, John D.** 1956. The Ixodoidea of Canada, Pub. 930, Canada Dept. Agr., Ottawa.
40. **Herms, W.B.** and **James, M.T.** 1966. Medical Entomology, New York: The Macmillan Co.
41. **Hite, Julia M.** et al. 1966. Biology of the Brown Recluse Spider, Bul. 711, Agr. Exp. Sta., Univ. of Arkansas, Fayetteville.
42. **Holland, George P.** 1949. The Siphonaptera of Canada, Canada Dept. Agr., Pub. 817, Tech. Bul. 70, Ottawa.
43. **Hubbard, C.A.** 1947. Fleas of Western North America: Their Relation to Public Health, reprint New York: Hafner Pub. Co. 1968.

44. **Hungerford, H.B.** 1933. "The Genus *Notonecta* of the World (Notonectidae: Hemiptera)", Univ. of Kansas Sci. Bul., 21(1):4-195.
45. **Hurd, Paul D., Jr.** and **Linsley, E.G.** 1964. "The Squash and Gourd Bees — Genera *Peponapis* Robertson and *Xenoglossa* Smith — Inhabiting North America North of Mexico (Hymenoptera: Apoidea)", Hilgardia, 35(15):375-477.
46. **Johansen, Carl.** 1966. Beekeeping, PNW Bul. 79, Washington State Univ., Pullman.
47. **Kaston, B.J.** 1970. Comparative Biology of American Black Widow Spiders, Trans. San Diego Soc. of Nat. History.
48. ——————1972. How to Know the Spiders, Dubuque, Iowa: Wm. C. Brown.
49. **Knipling, E.F.** 1952. "The Control of Insects Affecting Man", in Insects: Yearbook of Agriculture 1952, Washington, D.C.: Govt. Printing Office.
50. **Lavigne, R.A.** and **Fisser, H.G.** 1967. Controlling Western Harvester Ants, Mountain States Regional Pub. 3, Univ. of Wyoming, Laramie.
51. **Linsley, E. Gorton.** 1958. "The Ecology of Solitary Bees", Hilgardia, 27(19):543-599.
52. Los Angeles Times, June 7, 1969 (writer: Harry Nelson).
53. **Mallis, Arnold.** 1941. "A List of the Ants of California with Notes on Their Habits and Distribution", Bul. So. Calif. Acad. Sci., 40:61-100.
54. **Malloch, J.R.** 1914. American Black Flies or Buffalo Gnats, U.S. Dept. Agr., Bureau Ent., Tech. Series No. 26.
55. **McGregor, S.C.** and specialists. 1967. Beekeeping in the United States, Agr. Handb. No. 335, U.S. Dept. Agr., Washington, D.C.
56. **Meglitsch, Paul A.** 1972. Invertebrate Zoology, New York: Oxford Univ. Press.
57. **Michener, Charles D.** and **Michener, Mary H.** 1951. American Social Insects, New York: D. Van Nostrand.
58. ——————et al. 1972. Committee on the African Honey Bee, National Research Council, prepared for Agricultural Research Service, Washington, D.C. Distributed by National Technical Information Service, U.S. Dept. Commerce, Springfield, Va.
59. Military Entomology Operational Handbook. 1965. Dept. of Defense U.S.A.: Depts. of the Navy, The Army and The Airforce, Washington, D.C.
60. **Miller, C.D.F.** 1961. Taxonomy and Distribution of the Nearctic *Vespula*, Supp. 22, Canad. Ent.
61. **Milliron, H.E.** 1971, 1973. "A Monograph of the Western Hemisphere Bumblebees (Hymenoptera: Apidae: Bombinae)", Mem. Ent. Soc. of Canada, Nos. 82, 89, 91:1-333.
62. **Muma, Martin H.** 1967. "Scorpions, Whip Scorpions and Wind Scorpions", Arthropods of Florida, Vol. 4, Fla. Dept. Agr.
63. ——————1970. "A synoptic Review of North American, Central American, and West Indian Solpugida", *ibid.,* Vol. 5.
64. National Enquirer, Oct. 11, 1970 (writer: Walter Blaine).
65. **Neilson, C.L.** and **Gregson, J.D.,** 1967. Ticks and Man, Pub. 67-14, Dept. Agr., Province of British Columbia, Victoria.
66. **Off Belay.** 1972. Insect Repellents, June:49.
67. **Okumura, George T.** 1967. "A Report of Canthariasis and Allergy Caused by *Trogoderma* (Coleoptera: Dermestidae)", Calif. Vector Views, 14(3):19-24.
68. ——————1972. "Warehouse Beetle a Major Pest of Stored Food", Natl. Pest Control Oper. News, 32(1):4-5, 24.

69. **Petrunkevich, Alexander.** 1955. "The Spider and the Wasp", in First Book of Animals (A Scientific American Book), New York: Simon & Schuster.
70. **Pfeiffer, Ehrenfried.** 1949. Weeds and What They Tell, Emmaus, Pa.: Rodale Press.
71. **Pinto, John D.** and **Selander, Richard B.** 1970. The Bionomics of Blister Beetles of the Genus *Meloe* and a Classification of the New World Species, Illinois Biol. Monog. 42, Univ. of Illinois, Urbana.
72. **Pratt, Harry D.** Mites of Public Health Importance and Their Control (Developmental), Communicable Disease Center, U.S. Dept. of Health, Education, and Welfare, Atlanta, Ga.
73. ———1961. Lice of Public Health Importance and Their Control (Training Guide), Public Health Service Pub. No. 772, *ibid.*
74. ———and **Littig, Kent S.** 1962. Ticks of Public Health Importance and Their Control, Public Health Serv. Pub. No. 772, *ibid.*
75. ———and **Wiseman, John S.** Fleas of Public Health Importance and Their Control (Developmental), *ibid.*
76. **Robertson, R.L.** 1966. Stinging Caterpillars, Ext. Folder No. 251, Agr. Ext. Serv., North Carolina State Univ., Raleigh.
77. **Savory, T.H.** 1966. "False Scorpions", Scientific American, Mar.:95-101.
78. **Scheibner, R.A.** and **Knapp, F.W.** 1967. Wood Ticks, Leaf. 306, Coop. Ext. Serv., Univ. of Kentucky, Lexington.
79. **Scott, Harold G.** and **Pratt, Harry D.** 1959. Scorpions, Spiders, and Other Arthropods of Minor Public Health Importance and Their Control (Training Program), Communicable Disease Center, U.S. Dept. of Health, Education, and Welfare, Atlanta, Ga.
80. ———and **Littig, Kent S.** 1964. Flies of Public Health Importance and Their Control, Public Health Serv. Pub. No. 772, *ibid.*
81. **Silver, G.T.** 1958. "Studies on the Silver-spotted Tiger Moth, *Halisidota argentata* Pack. (Lepidoptera: Arctiidae) in British Columbia", Canad. Ent., 90(2):66.
82. **Smart, John.** 1965. A Handbook for the Identification of Insects of Medical Importance, The British Museum (Natural History), London.
83. **Smith, Leslie M.** and **Lowe, Homer.** 1948. "The Black Gnats of California", Hilgardia, 18(3):157-184.
84. **Smith, Marion R.** 1965. House-infesting Ants of Eastern United States: Their Recognition, Biology, and Economic Importance, Tech. Bul. No. 1326, U.S. Dept. Agr., Washington, D.C.
85. **Stahnke, Herbert L.** 1959. Scorpions, Tempe: Arizona State Univ.
86. **Stehr, F.W.** and **Cook, E.F.** 1968. A Revision of the Genus *Malacosoma* Hubner in North America (Lepidoptera: Lasiocampidae): Systematics, Biology, Immatures, and Parasites, U.S. Natl. Mus. Bul. 276.
87. **Stone, Alan.** 1930. "The Bionomics of Some Tabanidae (Diptera)", Ann. Ent. Soc. Amer., 23:261-304.
88. ———1938. The Horseflies of the Subfamily Tabaninae of the Nearctic Region, Misc. Pub. No. 305, U.S. Dept. Agr., Washington D.C.
89. **Swan, L.A.** and **Papp, C.S.** 1972. The Common Insects of North America, New York: Harper & Row.
90. **Taboada, Oscar.** 1967. Medical Entomology, Naval Medical School, National Naval Medical Center, Bethesda, Md.

91. Time Magazine. 1969. Infectious Diseases: Warning! Look to the Soles, Jan. 24:58. (Based on report by Edward S. Murray in the New England Journal of Medicine.)
92. **Truxal, Fred S.** 1953. "A Revision of the Genus *Buenoa* (Hemiptera: Notonectidae)", Univ. of Kansas Sci. Bul. 35, Pt. II:1351-1523.
93. United States Dept. of Agriculture. 1958. Wasps and How to Control Them, Leaf. No. 365.
94. ——————1955. Wood Ticks: How to Control Them in Infested Places, Leaf. 387.
95. ——————1966. Chiggers: How to Fight Them, Leaf. No. 403.
96. **Usinger, R.L.,** Ed. 1963. Aquatic Insects of California, Berkeley: Univ. of California Press.
97. **Vance, A.M.** and **App, B.A.** 1969. Lawn Insects: How to Control Them, Home and Garden Bul. No. 53, U.S. Dept. Agr., Washington, D.C.
98. **Wallis, J.B.** 1961. The Cicindelidae of Canada, Toronto: Univ. of Toronto Press.
99. **Walstrom, R.J.** and **Jones, P.A.** 1969. Alfalfa Leaf-Cutter Bee Management for Alfalfa Pollination in South Dakota, Bul. 544, Agr. Exp. Sta., South Dakota State Univ., Brookings.
100. **Wheeler, W.M.** 1910. Ants, Their Structure, Development, and Behavior, New York: Columbia Univ. Press.
101. **Williams, Stanley C.** 1970. A Systematic Revision of the Giant Hairy-Scorpion Genus *Hadrurus* (Scorpionida: Vejovidae), Occasional Papers No. 87, Calif. Acad. Sci.
102. **Wingo, Curtis W.** 1969. Poisonous Spiders and Other Venomous Arthropods in Missouri, Bul. 738, Agr. Exp. Sta., Univ. of Missouri, Columbia.
103. **Wirth, W.W.** 1952. The Helediae of California, Univ. of Calif. Publs. Ent., Vol. 9.
104. **Wood, Sherwin F.** 1951. "Bug Annoyance in the Sierra Nevada Foothills of California", Bul. So. Calif. Acad. Sci., 50(2):106-112.
105. ——————**,Mehringer, P.J.** and **Anderson, R.A.** 1961. "Conenose Bug Annoyance and *Trypanosoma cruzi* in Griffith Park in 1960", *ibid.,* 60(3):190-192.
106. **Young, Jerry H.** and **Coppock, Stanley.** 1966. The Texas Harvester Ant, OSU Ext. Facts NO. 7152, Oklahoma State Univ., Stillwater.

---

107. Check List of the Lepidoptera of America North of Mexico (including Greenland). — R.W Hodges, editor. — E.W. Classey Ltd. and the Wedge Entomological Research Foundation. — London 1983, I-XXIV + 283 pp.

INDEX TO ILLUSTRATIONS, COMMON NAMES, AND SCIENTIFIC NAMES

References in parentheses refer to illustrations.

Abedus 98
  identatus (40F)97, 98
Acanthomyops interjectus (20A)52
Acarina 125
Aedes (31)78
  aboriginis 82
  aegypti (33B)81
  cantator 82
  communis 82
  dorsalis 81, 82
  sollicitans 80, (34D)84
  squamiger 81
  sticticus 82
  taeniorhynchus 80, (34C)84
  vexans 82
Allodermanyssus
  sanguineus (53E)129, 130
Amblyomma 132, 139
  americanum (54E)133
  maculatum (54F)133
Andrea carlini (5E)18
Anopheles (31)78
  freeborni 83, (34F)84
  occidentalis 83
  quadrimaculatus (34E)84
Ant 46
  Argentine 48, (20E)52
  Army 48
  Big-headed (20H)52
  Carpenter 58
    Black (20B)52, (24B)58
    Brown (20C)52, (24A)58
    Giant 58
  Cornfield (20G)52
  Fire (19)49, 56
    Southern (21C)53
    Tropical (21E)53
  Harvester 50
    Black 51
    Red 61
    Texas 61
    Western (22)54, 61
  Imported fire (21F)53, 55, (23)57, 59, (26)60
  Legionary 48, (20D)52
  Little black (21B)53
  Odorous house (20F)52
  Pavement (21H)53
(Ants continued)
  Pharaoh (21A)53, 55
  Pyramid 51
  Southern fire 56, 61
  Texas leafcutting (21G)53
  Thief, Western 51, (21D)53
  Yellow (20A)52
Ant cricket 58
Anthophora occidentalis (5G)18
Aphonopelma (60)152, 153
  baileyi 154
  eutylenum 154
  reversum 154
Apis mellifera (3)14, (4)16, (9)24, (10, 11)25
Araneida 125
Arctia caja 118
Argas persicus 137
Arilus cristatus (39A)96
Atta texana (21G)53
Automeris io 111

B

Bachypelma (60)152
Backswimmer (39D)96
Bee wolf 40
Bembix comata 40
  spinolae (15B)38
Bees 15
  African 28
  Alkali 20, (6)20, (7)21
  Brazilian 28
  Bumble bees 11, (12B)30
    Nevada 12
    Yellow 12
    Yellow-faced 14
  Carpenter (5C, D, H)18
  Honey (3)14, 15, 25
  Leaf cutting (5A, B)18, 21, (12A)30
  Mason (5F)18
  Mining (5E, G)18
  Squash 17
  Sweat (5I)18
Beetles 101
  Blister 103, (42E, F)104
    Black 105
    Green 105
    Infernal 105

(Beetles, Blister continued)
Nuttall 105
Spotted 105
White spotted 105
Charcoal 103, (42A)104
Darkling 108, (44)108
Dung (42K, L)104
Flatheaded borers (42B, C)104
Giant soldier 106
Rhinocerus (42J)104
Stag (42D)104
Tiger 102, (41)102
California black 102, (42I)104
LeConte's black 102, (42G)104
Oregon (41)102, 103, (42H)104
Warehouse 106, (43)107
Belostoma 98
flumineum 98
Biting midges 63
Black flies 68
Striped 68
Blue bug 137
Bombus (12B)30
Bombylius major (29I)67
Brachypelma 152
Buenoa margaritacea 100
Bugs 95
Assassin 95, (39C)96
Bed (40C)97, 100
Conenose (39H)96
Giant water (40E)97, 98
Tropical bed 100
Water 98
Wheel 96
Bumble bees 11, (1)13
Nevada 12
Yellow 12
Yellow faced 14

C

Caddicefly (40H)97
Calliphora (28C)65
Camponotus
castaneus (20C)52, (24A)58
herculeanus modoc 59
laevigatus 58
pennsylvanicus (20B)52, (24B)58
sansabaenus maccooki 58
s. vicinus 58
Canthon laevis (42L)104
Caterpillars 110
Leaf slug (45D)112
Pus (45B)112, 119
Range 113, 122
Spiny elm 118
Tent 120
Western tent 120
Cediopsylla simplex 178, (1410)178
Centipedes 165, (66A)165
Centruroides
gracilis 162
marginatus (64D)160
sculpturatus 153, (64F)160, 162
vittatus (64E)160, 162
Ceratophyllus gallinae 180, (1415)179
niger (1414)179, 180
Chigger 127, 128, (52)128, (53A)129
Chigoe (36C)89, 90, (1421)181
Chlorion ichneumonea (15I)38, 40
Chalybion californicum (16B)39, 40
Chrysops 71
coquilletti 72
discalis (27F)64, 72
excitans 72
proclivis 72
Cicada killer (15C)38
Cicindela oregona (41)102, 103, (42H)104
scutellaris 103
Cimex hemipterus 100
lectularius (40C)97, 100
Coelocnemis magna 109
Conomyrma insana 51
Copris carolinus (42K)104
Crane fly (34G-I)84
Coratina acantha (5D)18
Cryptanura brachiformis (15F)38
Cryptoglossa laevis (44A)108
Culex (31)78
pipiens 83, (34A)84
quinquefasciatus 83
tarsalis 83
Culicoides 65, 66
furens (27E)64
obsoletus 66
reevesi 67
tristriatulus 66
Culiseta 78
alaskaensis 79
incidens 79
Ctenocephalides canis 88, (36B)89, 177, (1406)178
felis 88, (36A)89, 177, (1408)178
Ctenopsyllus 179

D

Dalcerides ingenitus 119
Dasymutilla bioculata (15J)38
gloriosa 41
sackeni 41
Deer flies 70, 72
Demodex canis 132
cati 132
equi 132
folliculorum (53F)129, 132
Dermacentor 132, 140
albipictus 135
andersoni (54B, Da)133
occidentalis (54Db)133

variabilis (54C)133, 135
Dermanyssus gallinae 130
Dermatobia hominis (29F)67
Diamanus montana (1417)179, 180
Dipylidium caninum 177
Dolichovespula 33
Dorymyrmex pyramicus 51
Dynastes tytius (42J)104

E

Earwig (40A, B)97
Echidnophaga gallinacea (36F)89, 90, 181, (1420)181
Eleodes 108
acuticaudus 109
armatus (44B, C)108
Epicauta maculata 105
pardalis 105
pennsylvanica 105
vittata (42F)104
Eremorhax magnus 149
striatus 149
Euborellia annulipes (40B)97
Eumenes bolli 36
b. oregonensis 36
crucifera 36
c. flavitinctus 36
c. nearcticus 36
fraterna (5J)18
Eutrombicula 127, 128

F

Fannia benjamini 74
canicularis 74
scalaris 74
Fleas 87, 175
American mouse 181, (1418)181
Bat 182, (1422)181
California ground squirrel (1417)179, 180
Cat 88, (36A)89, 177, (1408)178
Chicken (1415)179
Chipmunk 181, (1419)181
Dog 88, (36B)89, 177, (1406)178
European mouse 179, (1411)179
Gray squirrel (1416)179, 180
Human 90, (36D)89, 176, (1407)178
Northern rat 179, (1412)179
Oriental rat 90, (36E)89, 177, (1409)178
Pocket gopher 179, (1413)179
Rabbit (1410)178
Sand 182
Sticktight 90, (36F)89, 181, (1420)181
Western chicken (1414)179, 180
Flies 63
Bee (29I)67
Black (27A, D)64, (28A)65
Bot (29F)67
(Flies continued)
Bottle (28C)65
Deer (27F)64
False stable (27B)64, 73
Flower (29H)67
Horse (27G, H)64, (28B)65
House (29E)67, (30C)70, 73
Latrine 74
Lesser house 74
Little house 74
Louse (29G)67
Moth (27I)64
Nonbiting stable 73
Pigeon 77
Sand (27J)64
Snipe (27K)64, (30B)70, 75
Stable (29C)67, (30A)70, 73
False (29B)67
Forficula auricularia (40A)97
Foxella ignota 179, (1413)179

G

Gallinipper (33A)81
Geophilus californicus 166
Giant whipscorpion 147
Gloveria medusa 119
Glycyphagus (53C)129
Gnats 65, 76
Bodega black 65
Eye (29A)67, 76
Turkey 69
Valley black 65
Golden digger, great 40
Grapeleaf skeletonizer 118, (48)120, (49)121

H

Hadrurus (57)148, 163
arizonensis (64A)160, 163
spadix 163
Haemaphysalis 132
chordeilis 136
leporispalustris 136
Halictus farinosus 20
zonulum (5I)18
Halysidota argentata 117
caryae 117
Harrisina brillians 118, (48)120, (49)121
Hemerocampa leucostigma (46B)114, (47B)115, 116 (now *Orgyia)*
vetusta 116
Hemileuca 111, 113
burnsi 112
electra 113
juno 113
maia 111
oliviae 113
Hippelates 76

(*Hippelates* continued)
collusor 76
hermsi 76
pallipes 76
pusio (29A)67, 76
Hornets 33
Baldfaced 33
Giant 33
Horse flies 70
Black 71
California 71
Western 71
Hylaeus modestus (5F)18
Hymenolepis diminuta 178, 179
nana 177, 178

I

Iridomyrmex humilis 48, (20E)52
Ixodes 132, 136, 139
californicus 136
pacificus 136
ricinus 136
scapularis (54G)133

J

Jerusalem cricket (40D)97
Jigger 182

L

Labidus coecus (20D)52
Lasiohelea 65
Lasius alienus (20G)52
Latrodectus 154
bishopi 155
geometricus 154
hesperus 154, 155
mactans 154
variolus 154
Leptoconops 65, 86
kerteszi 65
torrens 65
Leptopsylla segnis 179, (1411)179
Lethocerus americanus (40E), 97, 98
Louse 92
Body 92, (38A)92, 93
Crab 92, (38C)92,
Head 92, (38B)92, 93
Pubic 92, 93
Louse flies 76
Loxosceles 154
arizonica 157
laeta 157
reclusa (62)156, 156
unicolor 157
Lynchia americana (29G)67
Lytta nuttalli (42E, F)104, 105
stygica 105
vesicatoria 105

M

Malacosoma 120
californicum 120
Mastigoproctus giganteus 147
Megabombus frevidus (1B)13
*f. californicus*12
nevadensis (1A)13
Megachile (12A)30
latimanus (5A)18, 22
rotundata (5B)18, (8)22
Megalopyge opercularis 119
Melanoplia acuminata 103, (42A)104
consputa 103
Meloe 105
Melophagus ovinus (29D)67, 77
Midges 63
Biting (27E)64
Millipedes (66B)165
Mischocyttarus flavitarsis
*flavitarsis*36
Mites (51B)126, 127
Cheese 130
Chicken 130
Chigger 130
Follicle (53F)129, 132
Fowl (54A)133
Grain 129
Grocer's itch 129, (53C)129
House mouse (53E)129, 130
Itch (53B)129, 131
Northern fowl 131
Pore (53F)129, 132
Straw itch (53G)129, 131
Tropical rat (53D)129, 130
Monobia quadridens (5N)18, 36
texana 36
Monomorium minimum (21B)53
pharaonis (21A)53, 55
Mosquitoes 77, (32)80
Alaska 79
Black saltmarsh 80
Brown saltmarsh 82
California saltmarsh 81
Common 83
Malaria (34E)84
Common snow 82
Cool weather 79
Dark rice-field 79
Floodwater 82
Florida Glades 79
Northern house 83, (34A, B)84
Northwest coast 82
Rainbarrel 79
Saltmarsh 80, (34C, D)84
Southern house 83
Vexatious 82
Western malaria 83, (34F)84
Yellow fever (33B)81

Moths 111
  Brown day 113
  Brown tail (46C)114, (47C)115
  Buck 111
  Burn's buck 112
  Gloveria 119
  Io 111
  Nevada buck 112
  Saddleback (45A)112, (50)122
  Sagebrush sheep 114
  Sheep 113
  Silkworm, giant 111
  Tiger 117
    Garden 118
    Silverspotted 117
  Tussock 116
    Hickory 117
    Western 116
    White-marked (45C)112, (46B)114, (47B)115, 116
Mourningcloak (46A)114, (47A)115, 118
Mud Daubers 39, (15H)38
  Blue (16B)39, 40
  Common 39
  Pipe organ (16A)39
Musca domestica (29E)67, (30C, D)70
Muscina assimilis (29B)67, 73
  stabulans (27B)64, 73
Myodopsylla insignis (1422)181, 182
Myrmecophila oregonensis 58

N

Neivamyrmex californicus 50
  nigrescens 49, 50
  opacithorax 49
Nepa (39F)96, 99
Nesopsyllus fasciatus 179, (1412)179
Nomia melanderi (6)20
Norape ovina (45D)112
Notonecta insulata (39D)96
  kirbyi 99
  undulata 99
  unifasciata 99
Nygmia phaeorrhoea (46C)114, (47C)115
Nymphalis antiopa (46A)114, (47A)115, 118

O

Odynerus cinnabarinus 37
  erythrogaster 37
Omus californicus 102, (42I)104
  lecontei 102, (42G)104
Opisocrostis bruneri 180
Orchopeas howardi (1416)179, 180
  wickhami 180
Orgyia see *Hemerocampa*
Ornithodorus 132, 137
  coriaceus 137
(*Ornithodorus* continued)
  turicata 138
Ornithonyssus
  bacoti (53D)129, 130
  sylvarium 131, (54A)133

P

Pahuello 137
Parasa chloris huachuca 119
Parnopes edwardsi (5L)18
Pasteurella pestis 90, 178
  tularensis 176
Pediculus humanus
  humanus 92, (38A)92
  capitis 92, (38B)92
Peponapis 17
  pruinosa 19
Pepsis attoni 39
  formosa (5M)18, 39
  mildei 37
Pheidole bicarinata
  vinelandica 48, (20H)52
Philanthus ventilabris 40, (15A)38
Phlebotomus (27J)64
Phthirus pubis 92, (38C)92
Pico 182
Pogonomyrmex barbatus 50, 51
  californicus 50
  occidentalis 50, 51
Polistes (12C)30, 31, (13)32, (17)43
  exclamans exclamans 35
  fuscatus aurifer 35
  hunteri californicus 35
Ponera ergatandria 55
  trigona opacior 55
Prosimulium hirtipes 70
Pseudohazis eglanterina 113
  hera 115
Pseudolynchia canariensis 77
Pseudomasaris vespoides 37
Pseudomyrmex barbatus 61
  occidentalis 61
  pallidus 55, 59
Psithyrus 14
  insularis 15
Psorophora 78
  ciliata (33A)81
Pulex irritans (36D)89, 90, 176, (1407)178
  simulans 177
Pyemotes ventricosus (53G)129, 131
Pyrobombus griseocollis (1D)13
  hunti (1C)13
  morrisoni (2)13
  vosnesenskii 14

R

Ranatra brevicollis (39E)96, 99
  fusca 99

quadridentata 99
Rasahus thoracicus 95, (39B)96
Reduvius personatus (39G)96
Rhipicephalus sanguineus 135, 136
Rickettsis typhi 90, 177
prowazeki 93

S

Sarcoptes scabiei (53B)129, 131
Saturnia mendocini 116
Sceliphron caementarium (15H)38, 39
Scloropendra heros 166
Scolia dubia (15G)38, 41
d. haematodes 41
d. monticola 41
Scorpions (57)148, 159, (65)161
Black (64D)160
Common striped 162
Deadly sculptured 162
Deadly striped 162
Hairy (64A)160, 163
Margarite 162
Mordant 163
Stripe-tail devil (64B)160, 162
Whip (58)149
Sheep ked (29D)67, 77
Sibine stimulea (50)122
Simulium (28A)65
arcticum 69
bivittatum 68
griseum 69
meridionale 69
trivittatum 69
vittatum (27D)64, 68
Sinea diadema (39C)96
Solenopsis geminata (19)49, (21E)53, 56,
invicta 56
molesta (21D)53
m. validiuscula 51
richteri (21F)53, 56
xyloni (21C)53, 56
Spanishfly 104
Sphecius speciosus (15C)38
Sphex urnarius (15E)38
Spiders 150
Black widow 154, (61)155
Brown recluse 156, (62)156, (63)157
Recluse 156, (62)156
Violin 156, (62)156
Western black widow 155
Stenopelmatus fuscus (40D)97
Stenoponia americana 181, (1418)181
Stomoxys (30A)70
calcitrans (29C)67, 73
Sucking lice 92
Symphoromyia (30B)70, 75
atripes (27K)64
limata 75
pachyceras 75
plagens 75
Syrphus torvus (29H)67

T

Tabanus (28B)65, 71
atratus 71
californicus 71
frontalis 71
lineola (27H)64
opacus 71
punctifer 71
rhombicus 71
septentrionalis 71
sonomensis 71
Tamiophila grandis (1419)181
Tapinota sessile (20F)52
Tarantula hawk (5M)18, 37
Tarantulas (60)152, 153
Tegrodera erosa 106
Tetramorium caespitum (21H)53
Thysanoptera (40G)97
Ticks (5I)126, 132
American dog (54C)133, 135, (56)135
Bird 137
Black-legged (54G)133
Rocky Mt. wood 132, (54B, Da)133, 134, (56)135
Brown dog (55A)134, 135
Castor bean 136
Deer 136
Ear 137
Fowl (55B)134, 137
Gulf Coast (54F)133
Lone star (54E)133
Pacific coast (54Db)133, 135
Rabbit 136
Relapsing fever 138
Soft (55C)134
Winter 135
Tipula cunctans (34G)84
paludosa (34I)84
trivittata (34H)84
Tique 182
Toe biter (40F)97, 98
Triatoma 95, 101
protracta 95
sanguisuga (39H)96
Trichoptera (40H)97
Thrips (40G)97
Trogoderma 106
glabrum 107
ornatum 107
variabile 106, (43A, B)107
Trombicula
alfreddugesi 127
belkini 127
[Eutrombicula] batatas 127, (53A)129
irritans 128, (52)128
Trypanosoma cruzi 101

Trypoxylon clavatum (15D)38
politum (16A)39
Tunga penetrans (36C)89, 90, (1421)181, 182
Tyrophagus casei 130
castellanii 130

U

Uroctonus mordax 163

V

Vejovis flavus (64C)160
spinigerus (64B)160, 162
Velvet ant 41
Veromessor andrei 51
pergandei 51
Vespa crabro germana 33, (14B)34
Vespula 33
arenaria 45
consobrina (14A)34
maculata 33, (14C)34
pennsylvanica (14D)34, 34
sulphurea 35
vulgaris 34

W

Wasps 29, (12C)30, (13)32
Cuckoo (5L)18
Digger (15G, I)38, 40, 41
Ichneumon (15F)38
Mason (5N)18, 36
Paper 35, (14)34
Potter (5J)18, 36
Sand (15B, E)38, 40
Spider 37
Western sand 40
Zebra paper 35
Waterscorpion, Western (39E, F)96, 99
Western bloodsucking conenose 95
Western corsair 95, (39B)96
Wheel bug 96, (39A)96
Whipscorpions 147

X

Xenoglossa 17, 19
Xenopsylla cheopis (36E)89, 90, 177, (1409)178
Xylocopa tabaniformis (5H)18
virginica (5C)18

Y

Yellowjacket 33
Common 34
Pennsylvania 34